The Theory of Differential Equations

The Theory of Differential Equations

Editor
Abby Enger

scitus
academics

The Theory of Differential Equations
Edited by **Abby Enger**

ISBN: 978-1-68117-651-2
Library of Congress Control Number: 2016946518

SCITUS Academics LLC,
616, Corporate Way, Suite 2, 4766,
Valley Cottage, NY 10989

www.scitusacademics.com

Printed in United States of America on Acid Free Paper

Preface

Differential equations have a remarkable ability to predict the world around us. They are used in a wide variety of disciplines, from biology, economics, physics, chemistry and engineering. They can describe exponential growth and decay, the population growth of species or the change in investment return over time. A differential equation is a mathematical equation that relates some function with its derivatives. In applications, the functions usually represent physical quantities, the derivatives represent their rates of change, and the equation defines a relationship between the two. For over 300 years, differential equations have served as an essential tool for describing and analyzing problems in many scientific disciplines. Differential equations can be divided into several types. Apart from describing the properties of the equation itself, these classes of differential equations can help inform the choice of approach to a solution. Commonly used distinctions include whether the equation is: Ordinary/Partial, Linear/Non-linear, and Homogeneous/Inhomogeneous. The theory of differential equations is closely related to the theory of difference equations, in which the coordinates assume only discrete values, and the relationship involves values of the unknown function or functions and values at nearby coordinates. Many methods to compute numerical solutions of differential equations or study the properties of differential equations involve approximation of the solution of a differential equation by the solution of a corresponding difference equation.

The Theory of Differential Equations provides an introduction to many of the important topics associated with ordinary differential equations. A unique introduction to the subject, reflecting different approaches to the integration of differential equations.

Table of Contents

CHAPTER 1

On Approximate Solutions of Second-Order Linear Partial Differential Equations

Yousry S. Hanna

National Research Institute of Astronomy and Geophysics, Helwan, Egypt

ABSTRACT

In this paper, a Chebyshev polynomial approximation for the solution of second-order partial differential equations with two variables and variable coefficients is given. Also, Chebyshev matrix is introduced. This method is based on taking the truncated Chebyshev expansions of the functions in the partial differential equations. Hence, the result matrix equation can be solved and approximate value of the unknown Chebyshev coefficients can be found.

KEYWORDS

Chebyshev Polynomial; Differential Equations; Polynomial Approximation

1. INTRODUCTION

Let the second-order partial differential equation be in the form [1, 2]

$$A(x,y)u_{xx}+B(x,y)u_{xy}+C(x,y)u_{yy}+D(x,y)u_{x}$$
$$+E(x,y)u_{y}+F(x,y)u=G(x,y) \tag{1.1}$$

We assume that it has a Chebyshev series solution in the form

$$\begin{aligned}u(x,y)&=\frac{1}{4}a_{0,0}T_{0,0}(x,y)+\frac{1}{2}a_{0,1}T_{0,1}(x,y)+\cdots\\&+\frac{1}{2}a_{1,0}T_{1,0}(x,y)+a_{1,1}T_{1,1}(x,y)+\cdots\\&+\frac{1}{2}a_{2,0}T_{2,0}(x,y)+a_{2,1}T_{2,1}(x,y)+\cdots\\&=\sum_{r=0}^{\infty}\sum_{s=0}^{\infty}a_{r,s}T_{r,s}(x,y)\end{aligned} \tag{1.2}$$

where Σ denotes a sum whose first term is halved. The unknown coefficients $a_{r,s};\ r=0,1,2,\cdots,\ s=0,1,2,\cdots$ can be determined by using so called Chebyshev matrix method.

2. CALCULATION OF CHEBYSHEV COEFFICIENTS

Let we have a function $u(x,y)$, $(x,y)\in[-1,1]$ and its nth derivatives with respect to x can be expanded in Chebyshev series

$$u(x,y)=\sum_{r=0}^{\infty}\sum_{s=0}^{\infty}a_{r,s}T_{r,s}(x,y)$$

And

$$u^{(n,0)}(x,y)=\sum_{r=0}^{\infty}\sum_{s=0}^{\infty}a_{r,s}^{(n,0)}T_{r,s}(x,y)$$

Respectively, where $a_{r,s}^{(n,0)}$ and $a_{r,s}$ are Chebyshev coefficients; clearly, $a_{r,s}^{(n,0)}=a_{r,s}$ and $u^{(0,0)}(x,y)=u(x,y)$. Then the recurrence relation between the coefficients of $u^{(n,0)}(x,y)$ and $u^{(n+1,0)}(x,y)$ is obtained as

$$2ra_{r,s}^{(n,0)}=a_{r-1,s}^{(n+1,0)}-a_{r+1,s}^{(n+1,0)},\ r\geq 1,\ s\geq 0 \tag{2.1}$$

From Equation (2.1), we can deduce the relations

$$2(r+1)a_{r+1,s}^{(n,0)}=a_{r,s}^{(n+1,0)}-a_{r+2,s}^{(n+1,0)},$$
$$2(r+3)a_{r+3,s}^{(n,0)}=a_{r+2,s}^{(n+1,0)}-a_{r+4,s}^{(n+1,0)},$$
$$2(r+5)a_{r+5,s}^{(n,0)}=a_{r+4,s}^{(n+1,0)}-a_{r+6,s}^{(n+1,0)},$$

And adding these side by side, we get

$$a_{r,s}^{(n+1,0)}=2\left[(r+1)a_{r+1,s}^{(n,0)}+(r+3)a_{r+3,s}^{(n,0)}+(r+5)a_{r+5,s}^{(n,0)}+\cdots\right]$$

Or

$$a_{r,s}^{(n+1,0)}=2\sum_{i=0}^{\infty}(r+2i+1)a_{r+2i+1,s}^{(n,0)} \tag{2.2}$$

Specially, we can express the coefficients $a_{r,s}^{(1,0)}$ and $a_{r,s}^{(2,0)}$, in terms of the $a_{r,s}$ by means of Equation (2.2), in the forms

$$a_{r,s}^{(1,0)} = 2\sum_{i=0}^{\infty}(r+2i+1)a_{r+2i+1,s}^{(0,0)} \tag{2.3}$$

And

$$a_{r,s}^{(2,0)} = 2\sum_{i=0}^{\infty}(r+2i+1)a_{r+2i+1,s}^{(1,0)}$$

Or

$$a_{r,s}^{(2,0)} = 4\sum_{i=0}^{\infty}i(r+i)(r+2i)a_{r+2i,s}^{(0,0)} \tag{2.4}$$

Now, let us take $r=0,1,2,\cdots,N$ and assume $a_{r,s}^{(0,0)} = a_{r,s}^{(1,0)} = \cdots = 0$ for $r>N$; then the system (2.2) can be transformed into the matrix form,

$$A^{(n+1,0)} = 2MA^{(n,0)}, \quad n=0,1,2,\cdots, \tag{2.5}$$

where M is given in [3].

$$A^{(n,0)} = \begin{bmatrix} \frac{1}{4}a_{0,0}^{(n,0)} & \frac{1}{2}a_{0,1}^{(n,0)} & \cdots & \frac{1}{2}a_{0,N}^{(n,0)} \\ \frac{1}{2}a_{1,0}^{(n,0)} & a_{1,1}^{(n,0)} & \cdots & a_{1,N}^{(n,0)} \\ \vdots & \vdots & \cdots & \vdots \\ \frac{1}{2}a_{N,0}^{(n,0)} & a_{N,1}^{(n,0)} & \cdots & a_{N,N}^{(n,0)} \end{bmatrix}$$

For $n=0,1,\cdots$, it follows from Equation (2.5) that

$$\begin{aligned} A^{(1,0)} &= 2MA^{(0,0)}, \\ A^{(2,0)} &= 2MA^{(1,0)} = 2^2 M^2 A^{(0,0)}, \\ A^{(3,0)} &= 2MA^{(2,0)} = 2^3 M^3 A^{(0,0)}, \\ &\vdots \\ A^{(n,0)} &= 2MA^{(n-1,0)} = 2^n M^n A^{(0,0)}, \\ &\vdots \end{aligned} \tag{2.6}$$

where clearly $A^{(0,0)} = A$.

Let us assume, in the range $[-1,1]$, that the nth derivatives of $u(x,y)$ with respect to y can be expanded in Chebyshev series

$$u^{(0,n)}(x,y) = \sum_{s=0}^{\infty} a_{r,s}^{(0,n)} T_{r,s}(x,y)$$

Respectively, where $a_{r,s}^{(0,n)}$ and $a_{r,s}$ are Chebyshev coefficients; clearly $a_{r,s}^{(0,0)} = a_{r,s}$ and $u^{(0,0)}(x,y) = u(x,y)$. Then the recurrence relation between the coefficients of $u^{(0,n)}(x,y)$ and $u^{(0,n+1)}(x,y)$ is obtained as

$$2sa_{r,s}^{(0,n)} = a_{r,s-1}^{(0,n+1)} - a_{r,s+1}^{(0,n+1)}, \ r \geq 0, \ s \geq 1 \tag{2.7}$$

From Equation (2.7), we can deduce the relations

$$\begin{aligned} 2(s+1)a_{r,s+1}^{(0,n)} &= a_{r,s}^{(0,n+1)} - a_{r,s+2}^{(0,n+1)}, \\ 2(s+3)a_{r,s+3}^{(0,n)} &= a_{r,s+2}^{(0,n+1)} - a_{r,s+4}^{(0,n+1)}, \\ 2(s+5)a_{r,s+5}^{(0,n)} &= a_{r,s+4}^{(0,n+1)} - a_{r,s+6}^{(0,n+1)}, \\ &\vdots \end{aligned}$$

and adding these side by side, we get

$$a_{r,s}^{(0,n+1)} = 2\left[(s+1)a_{r,s+1}^{(0,n)} + (s+3)a_{r,s+3}^{(0,n)} + (s+5)a_{r,s+5}^{(0,n)} + \cdots\right]$$

Or

$$a_{r,s}^{(0,n+1)} = 2\sum_{i=0}^{\infty}(s+2i+1)a_{r,s+2i+1}^{(0,n)} \tag{2.8}$$

Specially, we can express the coefficients $a_{r,s}^{(0,1)}$ and $a_{r,s}^{(0,2)}$ in terms of the $a_{r,s}$, by means of Equation (2.8), in the forms

$$a_{r,s}^{(0,1)} = 2\sum_{i=0}^{\infty}(s+2i+1)a_{r,s+2i+1}^{(0,0)} \tag{2.9}$$

And

$$a_{r,s}^{(0,2)} = 2\sum_{i=0}^{\infty}(s+2i+1)a_{r,s+2i+1}^{(0,1)}$$

or

$$a_{r,s}^{(0,2)} = 4\sum_{i=0}^{\infty}i(s+i)(s+2i)a_{r,s+2i}^{(0,0)} \tag{2.10}$$

Now, let us take $s = 0,1,\cdots,N$ and assume $a_{r,s}^{(0,0)} = a_{r,s}^{(0,1)} = \cdots = 0$ for $s > N$; then the system (2.8) can be transformed into the matrix form,

$$A_t^{(0,n+1)} = 2MA_t^{(0,n)},\ n = 0,1,2,\cdots, \tag{2.11}$$

where

$$A_t^{(0,n)} = \begin{bmatrix} \frac{1}{4}a_{0,0}^{(0,n)} & \frac{1}{2}a_{1,0}^{(0,n)} & \cdots & \frac{1}{2}a_{N,0}^{(0,n)} \\ \frac{1}{2}a_{0,1}^{(0,n)} & a_{1,1}^{(0,n)} & \cdots & a_{N,1}^{(0,n)} \\ \vdots & \vdots & \cdots & \vdots \\ \frac{1}{2}a_{0,N}^{(0,n)} & a_{1,N}^{(0,n)} & \cdots & a_{N,N}^{(0,n)} \end{bmatrix}$$

For $n = 0,1,\cdots,$ it follows from Equation (2.11) that

$$\begin{aligned} A_t^{(0,1)} &= 2MA_t^{(0,0)}, \\ A_t^{(0,2)} &= 2MA_t^{(0,1)} = 2^2 M^2 A_t^{(0,0)}, \\ A_t^{(0,3)} &= 2MA_t^{(0,2)} = 2^3 M^3 A_t^{(0,0)}, \\ &\vdots \\ A_t^{(0,n)} &= 2MA_t^{(0,n-1)} = 2^n M^n A_t^{(0,0)}, \\ &\vdots \end{aligned} \tag{2.12}$$

where clearly $A_t^{(0,0)} = A_t = \left[A^{(0,0)}\right]^t$. Furthermore, $A_t^{(n,n)}$ can be expressed as follows:

$$\begin{aligned} A^{(0,n)} &= 2^n M^n A_t, \\ A^{(n,n)} &= 2^n M^n A\left(2^n M^n\right)^t \end{aligned} \tag{2.13}$$

3. FUNDAMENTAL RELATIONS

Now consider Equation (1.1), where A, B, C, D, E, F and G are functions of x and y, or constant, defined in the range $[-1,1]$. Our purpose is to investigate the truncated Chebyshev series solution of Equation (1.1), under the given conditions, in the series form

$$u(x,y)=\sum_{r=0}^{N}\sum_{s=0}^{N}a_{r,s}T_{r,s}(x,y)$$

or in the matrix form

$$u(x,y)=T_x A T_y \tag{3.1}$$

where $a_{r,s}$, $s=0,1,2,\cdots,N$ $r=0,1,2,\cdots,N$, are the Chebyshev coefficients to be determined $T_{r,s}(x,y)$ are the bivariate Chebyshev polynomials defined in [4], and matrices T_x, T_y and A are defined by

$$T_x=\begin{bmatrix}T_o(x) & T_1(x) & \cdots & T_N(x)\end{bmatrix},$$

$$T_y=\begin{bmatrix}T_o(y) & T_1(y) & \cdots & T_N(y)\end{bmatrix}^t,$$

$$\begin{bmatrix}\frac{1}{4}a_{0,0} & \frac{1}{2}a_{1,0} & \cdots & \frac{1}{2}a_{N,0}\\ \frac{1}{2}a_{0,1} & a_{1,1} & \cdots & a_{N,1}\\ \vdots & \vdots & \cdots & \vdots\\ \frac{1}{2}a_{0,N} & a_{1,N} & \cdots & a_{N,N}\end{bmatrix}^t$$

To obtain the solution of Equation (1.1) in the form of Equation (3.1), first we must reduce Equation (1.1) to a differential Equation whose coefficients are polynomials [5]. For this purpose, we assume that the functions $A(x,y)$, $B(x,y)$, $C(x,y)$, $D(x,y)$, $E(x,y)$, $F(x,y)$, and $G(x,y)$ can be expressed in the form

$$A(x,y)=\sum_{n=0}^{N}\sum_{m=0}^{N}a_{n,m}x^{n}y^{m},$$

$$B(x,y)=\sum_{n=0}^{N}\sum_{m=0}^{N}b_{n,m}x^{n}y^{m},$$

$$C(x,y)=\sum_{n=0}^{N}\sum_{m=0}^{N}c_{n,m}x^{n}y^{m},$$

$$D(x,y)=\sum_{n=0}^{N}\sum_{m=0}^{N}d_{n,m}x^{n}y^{m},$$

$$E(x,y)=\sum_{n=0}^{N}\sum_{m=0}^{N}e_{n,m}x^{n}y^{m},$$

$$F(x,y)=\sum_{n=0}^{N}\sum_{m=0}^{N}f_{n,m}x^{n}y^{m},$$

$$G(x,y)=\sum_{n=0}^{N}\sum_{m=0}^{N}g_{n,m}x^{n}y^{m} \tag{3.2}$$

Which are Taylor polynomials at $(x,y)=(0,0)$. By using the expressions (3.2) in Equation (1.1), we get

$$\sum_{n=0}^{N}\sum_{m=0}^{N}\Big[a_{n,m}x^{n}y^{m}u_{xx}+b_{n,m}x^{n}y^{m}u_{xy}+c_{n,m}x^{n}y^{m}u_{yy}$$
$$+d_{n,m}x^{n}y^{m}u_{x}+e_{n,m}x^{n}y^{m}u_{y}+f_{n,m}x^{n}y^{m}u\Big]$$
$$=\sum_{n=0}^{N}\sum_{m=0}^{N}g_{n,m}x^{n}y^{m} \tag{3.3}$$

The Chebyshev expansions of terms

$$x^{n}y^{m}u^{(p,t)}\left(p=0,1,2;n=0,1,\cdots,N;m=0,1,\cdots,N\right)$$

in Equation (3.3) are obtained by means of the formulae

$$x^n u^{(p,0)}(x,y) = \sum_{r=0}^{N}\sum_{s=0}^{N}\sum_{i=0}^{n} 2^{-n}\binom{n}{i} a^{(p,0)}_{|r-n+2i|,s} T_{r,s}(x,y),$$

$$y^m u^{(p,0)}(x,y) = \sum_{r=0}^{N}\sum_{s=0}^{N}\sum_{j=0}^{m} 2^{-m}\binom{m}{j} a^{(p,0)}_{r,|s-m+2j|} T_{r,s}(x,y),$$

$$x^n y^m u^{(p,0)}(x,y)$$
$$= \sum_{r=0}^{N}\sum_{s=0}^{N}\sum_{i=0}^{n}\sum_{j=0}^{m} 2^{-n-m}\binom{n}{i}\binom{m}{j} a^{(p,0)}_{|r-n+2i|,|s-m+2j|} T_{r,s}(x,y),$$

$$x^n y^m u^{(0,t)}(x,y)$$
$$= \sum_{r=0}^{N}\sum_{s=0}^{N}\sum_{i=0}^{n}\sum_{j=0}^{m} 2^{-n-m}\binom{n}{i}\binom{m}{j} a^{(0,t)}_{|r-n+2i|,|s-m+2j|} T_{r,s}(x,y),$$

$$x^n y^m u^{(p,t)}(x,y)$$
$$= \sum_{r=0}^{N}\sum_{s=0}^{N}\sum_{i=0}^{n}\sum_{j=0}^{m} 2^{-n-m}\binom{n}{i}\binom{m}{j} a^{(p,t)}_{|r-n+2i|,|s-m+2j|} T_{r,s}(x,y), \quad (3.4)$$

4. MATRIX FORMS OF TERMS IN THE EQUATION

The matrix representation of Equation (3.4) can be given by

$$x^p y^v u^{(0,0)} = T_x M_p A (M_v)^t T_y \quad (4.1)$$

$$x^p y^v u^{(1,0)} = T_x M_p 2MA (M_v)^t T_y \quad (4.2)$$

$$x^p y^v u^{(0,1)} = T_x M_p A (M_v 2M)^t T_y \quad (4.3)$$

$$x^p y^v u^{(2,0)} = T_x M_p 2^2 M^2 A (M_v)^t T_y \tag{4.4}$$

$$x^p y^v u^{(0,2)} = T_x M_p A (M_v 2^2 M^2)^t T_y \tag{4.5}$$

$$x^p y^v u^{(1,1)} = T_x M_p 2MA (M_v 2M)^t T_y \tag{4.6}$$

$$\sum_{n=0}^{N} \sum_{m=0}^{N} g_{n,m} x^n y^m = G(x,y) = T_x S T_y \tag{4.7}$$

where $S = D^t \bar{G} D = [s_{i,j}]$ and $\bar{G} = [g_{i,j}]$, $i,j = 0,1,2,\cdots,N$.

And for $p = 0,1,2,\cdots,N$; $v = 0,1,2,\cdots,N$, $M_p = [m_{i,j}]$ ($I = 0,1,\cdots,N+1$ and $j = 0,1,\cdots,N+1$) is a matrix of size $(N+1)\times(N+1)$. The elements of M_p are given in [6].

Substituting the expressions (4.1)-(4.7) into Equation (3.3), and simplifying the result, we have the matrix equation

$$\sum_{n=0}^{N} \sum_{m=0}^{N} \Big[4a_{n,m} M_n M^2 A (M_m)^t + 4b_{n,m} M_n MA (M_m M^2)^t + 4c_{n,m} M_n A (M_m M^2)^t + + 2d_{n,m} M_n MA (M_m)^t + 2e_{n,m} M_n A (M_m M^2)^t + f_{n,m} M_n A (M_m)^t \Big] = S \tag{4.8}$$

Which corresponds to a system $(N+1)\times(N+1)$ algebraic equations for the unknown Chebyshev coefficients $a_{n,m}; n,m = 0,1,\cdots,N$. Briefly, we can assume that Equation (4.8) is given in the form

$$\sum_{\sigma=1}^{6} W_\sigma A Y_\sigma = S \tag{4.9}$$

where $W_\sigma = [w_{m,n}], Y_\sigma = [y_{m,n}],\ \sigma = 1,2,\cdots,6$

Matrix Equation (4.9) can be reduce to new matrix equation by making use of

$$x_i(N+1)+\ell j(N+1)+k = w_{i,j}y_{k,\ell}\ ;\ k = 0,1,2,\cdots,N,$$
$$j = 0,1,2,\cdots,N,\ \ell = 0,1,2,\cdots,N,\ i = 0,1,2,\cdots,N$$

Then the new matrix Equation (4.7) becomes

$$\sum_{\sigma=1}^{6} X_\sigma C = \bar{S} \tag{4.10}$$

Where

$$X_\sigma = [x_{z,q}],\ z,q = 0,1,2,\cdots,(N+1)(N+1),$$
$$\sigma = 1,2,\cdots,6$$

And

$$C = [a_{00}\quad a_{01}\quad \cdots\quad a_{0N}\quad a_{10}\quad a_{11}\quad \cdots\quad a_{1N}\quad \cdots\quad a_{N0}\quad a_{N1}\quad \cdots\quad a_{NN}]',$$

$$\bar{S} = [s_{00}\quad s_{01}\quad \cdots\quad s_{0N}\quad s_{10}\quad s_{11}\quad \cdots\quad s_{1N}\quad \cdots\quad s_{N0}\quad s_{N1}\quad \cdots\quad s_{NN}]'$$

5. MATRIX FORMS OF CONDITIONS

Let the conditions of Equation (1.1) be given by

$$u(x,-1)+u(x,0)+u(x,1)+u^{(0,1)}(x,-1)+u^{(0,1)}(x,0)+u^{(0,1)}(x,1) = f(x) \tag{5.1}$$

$$u(-1,y)+u(0,y)+u(1,y)+u^{(1,0)}(-1,y)+u^{(1,0)}(0,y)+u^{(1,0)}(1,y)=g(x) \tag{5.2}$$

$$u(-1,-1)+u(-1,0)+u(-1,1)+u(0,-1)+u(0,0)+u(0,1)+u(1,-1)+u(1,0)+u(1,1)=\lambda \tag{5.3}$$

where f is a function of x, g is a function of y and λ is constant.

Then, there are the following matrix forms at x = −1, 0, 1 and similar way for y = −1, 0, 1;

$$T_x(-1)=[1 \quad -1 \quad 1 \quad -1 \quad \cdots],$$
$$T_x(0)=[1 \quad 0 \quad -1 \quad 0 \quad 1 \quad \cdots],$$
$$T_x(1)=[1 \quad 1 \quad 1 \quad 1 \quad \cdots]$$

Derivative of T_x at x = −1, 0, 1 and similar way for y = −1, 0, 1;

$$T_x^{(1)}(-1)=\begin{bmatrix}0 & 1 & -2^2 & 3^2 & -4^2 & 5^2 & \cdots\end{bmatrix},$$
$$T_x^{(1)}(0)=[0 \quad 1 \quad 0 \quad -3 \quad 0 \quad 5 \quad 0 \quad -7 \quad \cdots],$$
$$T_x^{(1)}(1)=\begin{bmatrix}0 & 1^2 & 2^2 & 3^2 & 4^2 & 5^2 & \cdots\end{bmatrix}$$

We assume that the functions $f(x)$ and $g(y)$ can be expanded as

$$f(x)=\sum_{r=0}^{N} f_r T_r(x)$$

And

$$g(y)=\sum_{s=0}^{N} g_s T_s(y)$$

or in the matrix form

$$f(x) = T_x f$$

And

$$g(y) = g T_y$$

where

$$f = \begin{bmatrix} \frac{1}{2} f_0 & f_1 & \cdots & f_N \end{bmatrix}^t,$$

$$g = \begin{bmatrix} \frac{1}{2} g_0 & g_1 & \cdots & g_N \end{bmatrix}^t$$

In addition, at x = −1, 0, 1 and y = −1, 0, 1, we obtain the matrix forms

$$u(x,-1) = T_x \cdot A \cdot T_y(-1),$$
$$u(x,0) = T_x \cdot A \cdot T_y(0),$$
$$u(x,1) = T_x \cdot A \cdot T_y(1),$$

$$u^{(0,1)}(x,-1) = T_x \cdot A \cdot T_y^{(1)}(-1),$$
$$u^{(0,1)}(x,0) = T_x \cdot A \cdot T_y^{(1)}(0),$$
$$u^{(0,1)}(x,1) = T_x \cdot A \cdot T_y^{(1)}(1),$$

$$u(1,-1) = T_x(1) \cdot A \cdot T_y(-1),$$
$$u(-1,y) = T_x(-1) \cdot A \cdot T_y,$$
$$u(0,y) = T_x(0) \cdot A \cdot T_y,$$
$$u(1,y) = T_x(1) \cdot A \cdot T_y,$$

$$u^{(1,0)}(-1,y)=T_x^{(1)}(-1)\cdot A\cdot T_y,$$
$$u^{(1,0)}(0,y)=T_x^{(1)}(0)\cdot A\cdot T_y,$$
$$u^{(1,0)}(1,y)=T_x^{(1)}(1)\cdot A\cdot T_y,$$
$$u(1,0)=T_x^{(1)}(1)\cdot A\cdot T_y(0),$$
$$u(-1,-1)=T_x(-1)\cdot A\cdot T_y(-1),$$
$$u(-1,0)=T_x(-1)\cdot A\cdot T_y(0),$$
$$u(-1,1)=T_x(-1)\cdot A\cdot T_y(1),$$
$$u(0,-1)=T_x(0)\cdot A\cdot T_y(-1),$$
$$u(0,0)=T_x(0)\cdot A\cdot T_y(0),$$
$$u(0,1)=T_x(0)\cdot A\cdot T_y(1),$$

Substituting theses matrices forms into conditions (5.1)-(5.3), and then simplifying, we get the fundamental matrix equations of conditions as follows:

$$AU=f,VA=g \text{ and } QAZ=\lambda \tag{5.4}$$

where

$$U=T_y(-1)+T_y(0)+T_y(1)+T_y^{(1)}(-1)+T_y^{(1)}(0)+T_y^{(1)}(1)$$
$$V=T_x(-1)+T_x(0)+T_x(1)+T_x^{(1)}(-1)+T_x^{(1)}(0)+T_x^{(1)}(1)$$
$$Q=T_x(-1)+T_x(0)+T_x(1),$$
$$Z=T_y^{(1)}(-1)+T_y^{(1)}(0)+T_y^{(1)}(1)$$

6. FORMER METHOD FOR THE SOLUTION

We can assume that Equation (6.1) is in the form

$$\bar{X} \cdot C = \bar{S} \tag{6.1}$$

where $\bar{X} = \sum_{\sigma=1}^{6} X_\sigma$.

Then the augmented matrix of Equation (6.1) becomes $[\bar{X}; \bar{S}]$ or

$$\begin{bmatrix} \overline{x_{0,0}} & \overline{x_{0,1}} & \cdots & \overline{x_{0,N(N+2)}}; & s_{0,0} \\ \overline{x_{1,0}} & \overline{x_{1,1}} & \cdots & \overline{x_{1,N(N+2)}}; & s_{0,1} \\ \vdots & \vdots & \cdots & \vdots & \vdots \\ \overline{x_{N(N+2),0}} & \overline{x_{N(N+2),1}} & \cdots & \overline{x_{N(N+2),N(N+2)}}; & s_{N,N} \end{bmatrix} \tag{6.2}$$

If we take the new matrix forms of the conditions as $\bar{U}C = f$, $\bar{V}C = g'$ and $\bar{Q}C = \lambda$, respectively, the augmented matrices of them become $[\bar{U}; f]$, $[\bar{V}; g']$ and $[\bar{Q}; \lambda]$ or more clearly

$$\begin{bmatrix} \overline{u_{0,0}} & \overline{u_{0,1}} & \cdots & \overline{u_{0,N(N+2)}}; & \frac{1}{2} f_0 \\ \overline{u_{1,0}} & \overline{u_{1,1}} & \cdots & \overline{u_{1,N(N+2)}}; & f_1 \\ \vdots & \vdots & \cdots & \vdots & \vdots \\ \overline{u_{N,0}} & \overline{u_{N,1}} & \cdots & \overline{u_{N,N(N+2)}}; & f_N \end{bmatrix} \tag{6.3}$$

$$\begin{bmatrix} \overline{v_{0,0}} & \overline{v_{0,1}} & \cdots & \overline{v_{0,N(N+2)}}; & \frac{1}{2} g_0 \\ \overline{v_{1,0}} & \overline{v_{1,1}} & \cdots & \overline{v_{1,N(N+2)}}; & g_1 \\ \vdots & \vdots & \cdots & \vdots & \vdots \\ \overline{v_{N,0}} & \overline{v_{N,1}} & \cdots & \overline{v_{N,N(N+2)}}; & g_N \end{bmatrix} \tag{6.4}$$

And

$$\left[\overline{q_{0,0}} \quad \overline{q_{0,1}} \quad \cdots \quad \overline{q_{0,N(N+2)}}; \quad \lambda\right] \tag{6.5}$$

Consequently, by replacing Equations (6.3)-(6.5) by the last 2N + 1 rows of Equation (6.2), we have the new augment matrix

$$\begin{bmatrix} \overline{x_{0,0}} & \overline{x_{0,1}} & \cdots & \overline{x_{0,N(N+2)}}; & s_{0,0} \\ \overline{x_{1,0}} & \overline{x_{1,1}} & \cdots & \overline{x_{1,N(N+2)}}; & s_{0,1} \\ \vdots & \vdots & \cdots & \vdots & \vdots \\ \overline{x_{N^2-1,0}} & \overline{x_{N^2-1,1}} & \cdots & \overline{x_{N^2-1,N(N+2)}}; & s_{N-2,N-1} \\ \overline{u_{0,0}} & \overline{u_{0,1}} & \cdots & \overline{u_{0,N(N+2)}}; & \frac{1}{2}f_0 \\ \vdots & \vdots & \cdots & \vdots & \vdots \\ \overline{u_{N,0}} & \overline{u_{N,1}} & \cdots & \overline{u_{N,N(N+2)}}; & f_N \\ \overline{v_{0,0}} & \overline{v_{0,1}} & \cdots & \overline{u_{0,N(N+2)}}; & \frac{1}{2}g_0 \\ \vdots & \vdots & \cdots & \vdots & \vdots \\ \overline{v_{N,0}} & \overline{v_{N,1}} & \cdots & \overline{v_{N,N(N+2)}}; & g_N \\ \overline{q_{0,0}} & \overline{q_{0,1}} & \cdots & q_{N,N(N+2)}; & \lambda \end{bmatrix}$$

From the solution of this system we can find matrix C or matrix A.

7. APPLICATIONS

The Chebyshev matrix method applied in this study is useful in finding approximate solutions of second-order linear partial differential equations in both homogeneous and non-homogeneous cases, in terms of Chebyshev polynomials. We illustrate it by the following examples.

Example 1. We now consider the problem [7]:

$$u_{tt} = u_{xx} + 6;$$
$$u(x,0) = x^2,$$
$$u_t(x,0) = 4x \tag{1}$$

And seek the solution in the form

$$u_t(x,t) = \sum_{r=0}^{2}\sum_{s=0}^{2} a_{r,s}T_{r,s}(x,t),$$
$$(x,t) \in [-1,1] \tag{2}$$

Then we obtain the matrix equation

$$A\left(2^2 M^2\right)^t - 2^2 M^2 A = 6R, \tag{3}$$

Where

$$R = \begin{bmatrix} 1 & 0 & 0 \\ 0 & 0 & 0 \\ 0 & 0 & 0 \end{bmatrix}$$

And the condition matrices are

$$A \cdot T_t(0) = \begin{bmatrix} \frac{1}{2} & 0 & \frac{1}{2} \end{bmatrix}^t, \tag{4}$$

$$A \cdot T_t^{(1)}(0) = \begin{bmatrix} 0 & 4 & 0 \end{bmatrix}^t, \tag{5}$$

By replacing the new matrix form of Equations (4) and (5) in the new matrix form of Equation (3), we have the matrix equation under given conditions as follows:

$$\begin{bmatrix} 0 & 0 & 4 & 0 & 0 & 0 & -4 & 0 & 0 \\ 0 & 0 & 0 & 0 & 0 & 0 & 0 & -4 & 0 \\ 0 & 0 & 0 & 0 & 0 & 0 & 0 & 0 & -4 \\ 1 & 0 & -1 & 0 & 0 & 0 & 0 & 0 & 0 \\ 0 & 0 & 0 & 1 & 0 & -1 & 0 & 0 & 0 \\ 0 & 0 & 0 & 0 & 0 & 0 & 1 & 0 & -1 \\ 0 & 1 & 0 & 0 & 0 & 0 & 0 & 0 & 0 \\ 0 & 0 & 0 & 0 & 1 & 0 & 0 & 0 & 0 \\ 0 & 0 & 0 & 0 & 0 & 0 & 0 & 1 & 0 \end{bmatrix} \begin{bmatrix} \frac{1}{4}a_{0,0} \\ \frac{1}{2}a_{0,1} \\ \frac{1}{2}a_{0,2} \\ \frac{1}{2}a_{1,0} \\ a_{1,1} \\ a_{1,2} \\ \frac{1}{2}a_{2,0} \\ a_{2,1} \\ a_{2,2} \end{bmatrix} = \begin{bmatrix} 6 \\ 0 \\ 0 \\ \frac{1}{2} \\ 0 \\ \frac{1}{2} \\ 0 \\ 4 \\ 0 \end{bmatrix}$$

Hence, we obtain the augmented matrix

$$\begin{bmatrix} 0 & 0 & 4 & 0 & 0 & 0 & -4 & 0 & 0; & 6 \\ 0 & 0 & 0 & 0 & 0 & 0 & 0 & -4 & 0; & 0 \\ 0 & 0 & 0 & 0 & 0 & 0 & 0 & 0 & -4; & 0 \\ 1 & 0 & -1 & 0 & 0 & 0 & 0 & 0 & 0; & \frac{1}{2} \\ 0 & 0 & 0 & 1 & 0 & -1 & 0 & 0 & 0; & 0 \\ 0 & 0 & 0 & 0 & 0 & 0 & 1 & 0 & -1; & \frac{1}{2} \\ 0 & 1 & 0 & 0 & 0 & 0 & 0 & 0 & 0; & 0 \\ 0 & 0 & 0 & 0 & 1 & 0 & 0 & 0 & 0; & 4 \\ 0 & 0 & 0 & 0 & 0 & 0 & 0 & 1 & 0; & 0 \end{bmatrix}$$

The solution of this system is

$$A = \begin{bmatrix} \frac{5}{2} & 0 & 2 \\ 0 & 4 & 0 \\ \frac{1}{2} & 0 & 0 \end{bmatrix}$$

and thereby the solution of the problem (1) becomes

$$u(x,t) = \frac{5}{2} T_{0,0}(x,t) + 2T_{0,2}(x,t) + 4T_{1,1}(x,t) + \frac{1}{2} T_{2,0}(x,t)$$

or

$$u(x,t) = (x+2t)^2$$

This is exact solution [7].

Example 2

Let us now study the equation

$$x^2 u_{xx} - y^2 u_{yy} = 0$$

with conditions which are

$$u(0,y) = 1, \quad u(1,y) = e^y,$$
$$u(x,0) = 1, \quad u(x,1) = e^x,$$

The first four terms of the series expansions:

$$u(0,y)=1,\quad u(1,y)=e^{y}=1+y+\frac{y^2}{2}+\frac{y^3}{6},$$

$$u(x,0)=1,\quad u(x,1)=e^{x}=1+x+\frac{x^2}{2}+\frac{x^3}{6},$$

Chebyshev matrix forms of the conditions,

$$\begin{bmatrix}1 & 0 & -1 & 0 & 1 & 0\end{bmatrix}A=\begin{bmatrix}1 & 0 & 0 & 0 & 0 & 0\end{bmatrix}$$

$$\begin{bmatrix}1 & 1 & 1 & 1 & 1 & 1\end{bmatrix}A=\begin{bmatrix}\frac{5}{4} & \frac{9}{8} & \frac{1}{4} & \frac{1}{24} & 0 & 0\end{bmatrix}$$

$$A\begin{bmatrix}1\\0\\-1\\0\\1\\0\end{bmatrix}=\begin{bmatrix}1\\0\\0\\0\\0\\0\end{bmatrix},\quad A\begin{bmatrix}1\\1\\1\\1\\1\\1\end{bmatrix}=\begin{bmatrix}\frac{5}{4}\\\frac{9}{8}\\\frac{1}{4}\\\frac{1}{24}\\0\\0\end{bmatrix}$$

Matrix form of the equation is

$$M_2 4M^2A-A\left(M_2 4M^2\right)^t=0$$

$$\begin{bmatrix} 0 & 0 & \frac{1}{2} & 0 & 7 & 0 \\ 0 & 0 & 0 & \frac{9}{2} & 0 & \frac{55}{2} \\ 0 & 0 & \frac{1}{2} & 0 & 10 & 0 \\ 0 & 0 & 0 & \frac{3}{2} & 0 & \frac{35}{2} \\ 0 & 0 & 0 & 0 & 3 & 0 \\ 0 & 0 & 0 & 0 & 0 & 5 \end{bmatrix} A$$

$$-A\begin{bmatrix} 0 & 0 & 0 & 0 & 0 & 0 \\ 0 & 0 & 0 & 0 & 0 & 0 \\ \frac{1}{2} & 0 & \frac{1}{2} & 0 & 0 & 0 \\ 0 & \frac{9}{2} & 0 & \frac{3}{2} & 0 & 0 \\ 7 & 0 & 10 & 0 & 3 & 0 \\ 0 & \frac{55}{2} & 0 & \frac{35}{2} & 0 & 5 \end{bmatrix} = 0$$

From the solution of this matrix equation under the given conditions, we get the Chebyshev coefficients matrix as

$$A = \begin{bmatrix} \frac{9}{8} & 0 & \frac{1}{8} & 0 & 0 & 0 \\ 0 & \frac{35}{32} & 0 & \frac{1}{32} & 0 & 0 \\ \frac{1}{8} & 0 & \frac{1}{8} & 0 & 0 & 0 \\ 0 & \frac{1}{32} & 0 & \frac{1}{96} & 0 & 0 \\ 0 & 0 & 0 & 0 & 0 & 0 \\ 0 & 0 & 0 & 0 & 0 & 0 \end{bmatrix}$$

The solution of problem is obtained as

$$u(x,y)=1+xy+\frac{(x,y)^2}{2}+\frac{(x,y)^3}{6},$$

Which is the first four terms of $e^{(xy)}$.

8. CONCLUSIONS

Analytic solutions of the second order linear partial differential equations with variable coefficients are usually difficult. In many cases, it is required to approximate solutions. For this purpose, the Chebyshev matrix method can be proposed.

In this study, the usefulness of the Chebyshev matrix method presented for the approximate solution of the second order linear partial differential equations is discussed. Also, the method can be applied to both the nonhomogeneous and homogeneous cases.

A considerable advantage of the method is that the solution is expressed as a truncated Chebyshev series and thereby a Taylor polynomial. Furthermore, after calculation of the series coefficients, the solution $u(x,y)$ of the equations can be easily evaluated for arbitrary values of (x,y) at low computation effort.

An interesting feature of the Chebyshev matrix method is that the method can be used in finding exact solutions in much cases. The method can be also extending to the solution of the higher order linear partial differential equations.

REFERENCES

1. X. F. Yang, Y. X. Liu and S. Bai, "A Numerical Solution of Second-Order Linear Partial Differential Equations by Differential Transform," Journal of Applied Mathematics and Computing, Vol. 173, No. 2, 2006, pp. 792-802.

2. L. H. Ye and Z. T. Xu, "Oscillation Criteria for Second Order Quasilinear Neutral Delay Differential Equations," Journal of Applied Mathematics and Computing, Vol. 207, No. 2, 2009, pp. 388-396.

3. M. Sezer and M. Kaynak, "Chebyshev Polynomial Solutions of Linear Differential Equations," International Journal of Mathematical Education in Science & Technology, Vol. 27, No. 4, 1996, pp. 607-618.

4. N. K. Basu, "On Double Chebyshev Series Approximation," SIAM Journal on Numerical Analysis, Vol. 10, No. 3, 1973, pp. 496-505.

5. F. Cases, "Solution of Linear Partial Differential Equations by Lie Algebraic Methods," Journal of Computational and Applied Mathematics, Vol. 76, No. 1-2, 1996, pp. 159-170.

6. H. Koroglu, "Chebyshev Series Solution of Linear Fredholm Integrodifferential Equations," International Journal of Mathematical Education in Science & Technology, Vol. 29, No. 4, 1998, pp. 489-500.

7. V. S. Vladimirov, "A Collection of Problems on the Equations of Mathematical Physics," Mir Publishers, Moscow, 1986, pp. 139-149.

CHAPTER 2

An Algorithm to Optimize the Calculation of the Fourth Order Runge-Kutta Method Applied to the Numerical Integration of Kinetics Coupled Differential Equations

Sadao Isotani[1*], Walter Maigon Pontuschka[1], Seiji Isotani[2]

[1] Institute of Physics, University of São Paulo, São Paulo, Brazil

[2] Institute of Mathematics and Computer Science,
University of São Paulo, São Carlos, Brazil

ABSTRACT

The kinetic electron trapping process in a shallow defect state and its subsequent thermalor photo-stimulated promotion to a conduction band, followed by recombination in another defect, was described by Adirovitch using coupled rate differential equations. The solution for these equations has been frequently computed using the Runge-Kutta method. In this research, we empirically demonstrated that using the Runge-Kutta Fourth Order method may lead to incorrect and ramified results if the numbers of steps to achieve the solutions is not “large enough”. Taking into account these results, we conducted numerical analysis and experiments to develop an algorithm that determines the smallest non-critical number of steps in an automatic way to optimize the application of the Runge-Kutta Fourth Order method. This algorithm was implemented and tested in a variety of situations and the results have shown that our solution is robust in dealing with different equations and parameters.

KEYWORDS

Coupled Differential Equations; Runge-Kutta; Kinetics Equations

1. INTRODUCTION

Adirovitch proposed a description of the luminescence in crystal phosphor through a kinetic process involving electrons trapped in a crystal defect, their subsequent thermally or optically induced promotion to a conduction band, and their recombination in another defect [1]. The use of this model allows the inclusion of physically meaningful quantities [2,3]. To date, the model has been applied and further extended to analyse, for example, to the optical absorption decay of Fe impurity oxidation [4], the thermal decay of the optical absorption of irradiation-induced color centers in spodumene [5,6], the luminescence of irradiated spodumene [7], the EPR of atomic hydrogen in glass and beryl [8-10], defects in diamond films [11], charge transport in thin-layer photodielectric systems [12], thermally stimulated processes in dielectrics [13], stimulated luminescence in insulators and semiconductors [14], soliton in biological systems [15], photo-induced-degradation of amorphous hydrogen silicon (a-Si:H) [16] and the degradation of a polymer [17].

In Adirovitch's model, the concentration of trapped charges (Y_1) and free charges in the conduction band (Y_2) are related by the rate equations:

$$\frac{dY_1}{dT} = -P_1Y_1 + P_2Y_2(Y_0 - Y_1) \tag{1a}$$

$$\frac{dY_2}{dT} = P_1Y_1 - P_2Y_2(Y_0 - Y_1) - P_3Y_2(Y_1 + Y_2) \tag{1b}$$

where P_1, P_2 and P_3 are positive adjustable parameters associated with the release of charges from traps, retrapping and the charge and electron-hole recombinetion process, respectively. In general the energy dissipation of the recombination process occurs through a radiation process. Thus, the function $Y_2(Y_1 + Y_2)$ presented in Equation (1) is proportional to the luminescence in crystal phosphor and Y_0 is the concentration of traps [1].

Previous research findings have shown that the solutions for the non-linear stability analysis of equations 1a and 1b are stable, following a hyperbolic path [18]. Mizukami et al. [19] observed that these solutions can be represented by the sum of two terms. The first term leads to slow first order decay and the other to a fast decay. Based on these observations they have proposed an approximate method to reach stable solutions. To obtain the exact solutions, iterative methods can be applied. A family of explicit and implicit iterative methods that has been frequently used are the Runge-Kutta methods [20-22].

In this paper we initially report numerical analyses of the fourth order Runge-Kutta method as applied to the solution of Adirovitch model Equations (1a) and (1b). Even though the RK method is stable, we identified a disconcerting property that emerges from the stiffness of the method when solving these equations: the numerical results are reliable and efficient only for a limited range of parameter values. To solve this problem, this paper proposes an algorithm to determine the best values for parameters in the Runge-Kutta method in order to guarantee the reliability of the results while using the smallest numbers of steps to reach a solution. This algorithm was implemented and has been used in different situations with different parameters. Thus far, to the best of our knowledge, the proposed algorithm has not produced any incorrect solutions.

The paper is organized as follows: in the first section we present the numerical analysis of the Adirovitch model; next, we present the problem of calculating the smallest number of steps to find the solution of this model using a Runge-Kutta Method; and finally we present an algorithm, together with its implementation and tests, to optimize the calculation of the smallest noncritical number of steps to reach the solution.

2. NUMERICAL ANALYSIS

To numerically solve the Equations (1a) and (1b) for the Adirovitch model presented in the introduction of this work it is necessary to find the correct values for each adjustable parameter P_1, P_2, and P_3. Powerful genetic [23] and statistical [24] methods were developed recently and can be applied in fitting these parameters. Also, other classical methods for fitting can be adopted [25,26]. Nevertheless, fit methods may use a high number of iterations, and the number of steps used for the integration method may increase the computational cost. They may also require large adjustable parameter values. In this context, the solution of Equations (1) using the RungeKutta method, according to the parameters given, may require a large number of steps. Furthermore, the stiffness of the method can create difficulties in the numerical integration of Equations (1) resulting in abnormal and numerically unstable solutions. In this section we will examine the chaotic behaviours of these solutions. In the next section, we propose an algorithm to overcome them.

Equations (1) are made more tractable by introducing the transformations:

$$y_i = Y_i / Y_{1,0} \tag{2}$$

where $Y_{1,0} = Y_1(0)$ is the total concentration of trapped charges at the initial instant $T_0 = 0$, and $y_1(0) = 1$. Since the parameters P_1, P_2 and P_3 enter linearly in the right hand side of Equations (1), mathematically only two of them are really independent. Therefore, it is natural to make an additional scaling. Then, using $p_1 = 1,\quad p_2 = P_2/P_1, p_3 = P_3/P_1, t = P_1T$, we deal with only two parameters. Thus, Equations (1) are written as:

$$\frac{dy_1}{dt} = -y_1 + p_2 y_2 (y_0 - y_1) \tag{3a}$$

$$\frac{dy_2}{dt} = y_1 - p_2 y_2 (y_0 - y_1) - p_3 y_2 (y_1 + y_2) \tag{3b}$$

Numerical methods for the solution of Equations (3) can be divided into two categories [27]. One is the initial value problem, in which the values of $y_i(t)$ are known at an initial time t_0, with $y_i^0 = y_i(t_0)$, requiring the evaluation of $y_i(t)$. The other is the boundary condition for two points, in which boundary conditions are given for t_0 and t_f, typically.

The numerical solution of Equations (3) belonging to the first category are mostly grouped as Runge-Kutta [21, 22,27-30], Bulirsch-Stoer [31], predictor-corrector [32] solutions among others [33,34].

A widely used form of the Runge-Kutta method is of the fourth order. Using a vector notation typical in differential equations, $\boldsymbol{y}=(y_1,y_2)$ and $\boldsymbol{f}=(\mathrm{d}y_1/\mathrm{d}t,\mathrm{d}y_2/\mathrm{d}t)$, the advancing formula $\boldsymbol{y}^n \to \boldsymbol{y}^{n+1}$ from time t_n to $t_{n+1}=t_n+h$ is given by:

$$\boldsymbol{y}^{n+1}=\boldsymbol{y}^n+(1/6)\left(\boldsymbol{k}^1+2\boldsymbol{k}^2+2\boldsymbol{k}^3+\boldsymbol{k}^4\right)+\boldsymbol{O}^n \tag{4}$$

where $h=(t_{n+1}-t_n)/N$ is the time increment for N steps in which the interval is divided,

$$\begin{aligned}
\boldsymbol{k}^1 &= h\boldsymbol{f}\left(t_n;\boldsymbol{y}^n\right)\\
\boldsymbol{k}^2 &= h\boldsymbol{f}\left(t_n+h/2;\boldsymbol{y}^n+\boldsymbol{k}^1/2\right)\\
\boldsymbol{k}^3 &= h\boldsymbol{f}\left(t_n+h/2;\boldsymbol{y}^n+\boldsymbol{k}^2/2\right)\\
\boldsymbol{k}^4 &= h\boldsymbol{f}\left(t_n+h;\boldsymbol{y}^n+\boldsymbol{k}^3\right)
\end{aligned}$$

The precision of $\boldsymbol{O}^n$ is proportional to h^5. For orders greater than four, it is necessary to evaluate M to M + 2 more functions, increasing the computational cost. The equilibrium between computational efficiency and cost has been found for the fourth order Runge-Kutta method [13,23].

As the error is proportional to h^5, the Runge-Kutta method has a strong dependence on the number of steps. It predicts that the solution is invariant for large numbers of steps and that errors accumulate with any decrease in the number of steps.

Because of the dynamic nature of the Equations (3), it is possible that the accumulation of error makes the solutions erratic. Therefore, we investigated the types of errors that can happen depending on the number of steps and the values of the parameters of the Equations (3).

We have found at least four types of erratic solutions for Equations (3), which are shown in **Figure 1**. Here we assume that all charge traps are filled ($y_1(0)=1$) and have no free charges at first ($y_2(0)=0$). The time interval used for all calculations in the present work is $t\in[0,10]$. **Figure 1**(a) shows an error in which the solutions behave as expected for the exact solutions but whose values are wrong. **Figure 1**(b) shows deformation errors in which the solution has anomalous depressions or elevations. **Figure 1**(c) shows an error in which solutions are duplicated, suggesting chaotic behavior. Finally, **Figure 1**(d) shows erratic solutions leading to divergence. Erratic solutions such as divergence, doubling and deformations are easily identified. Nevertheless, wrong solutions are more difficult to identify.

The existence of incorrect solutions can be evaluated by analysing the correction for the n-th step given by $y_1^{n+1}-y_1^n$. **Figure 2** shows $y_1^{n+1}-y_1^n$ as a function of p_3 in the interval $p_3\in[0,80]$, with $p_2=65$, initial parameters $y_1(0)=1$, $y_2(0)=0$ and (a) $N=270$, (b) $N=230$, (c) $N=200$, and (d) $N=170$. Solid lines linking the values obtained at each i-th step makes the visualization of the properties of $y_1^{n+1}-y_1^n$ easy.

In **Figure 2**(a) ($N=270$) it is possible to see that the modulus of differences decreases at each step and grows slowly with the increase of p_3. This behaviour changes starting at $p_3\approx 75$, where oscillations can be seen. In **Figure 2**(b) ($N = 230$) a stranger fin-like formation appears at approximately $p_3 \approx 56$, at the same time that the differences become erratic. In **Figure 2**(c) ($N = 200$) the fin-like formation appears around $p_3 \approx 42$ and doubling appears in $p_3 \approx 68$. In the interval between the beginning of the fin-like formation and the beginning of the doubling, the solutions are incorrect. In **Figure 2**(d) ($N = 170$), we observed three critical points, one at the fin-like formation ($p_3 \approx 32$), the second at the doubling ($p_3 \approx 52$), and the other at a point where the solution diverges completely ($p_3 \approx 70$). The differences show the appearance of incorrect fin-like, doubling, and divergence type solutions, but they do not allow for prediction of the result of the sum of these differences. In fact we verified through an exact numerical solution of Equations (3) that wrong solutions for N = 170, 200, 230 and 270 begin at approximately $p_3 \approx 9, 21, 56, 75$, respectively. Therefore, $y_1^{n+1} - y_1^n$ is not capable of revealing cumulative errors and a new algorithm must be considered in order to do so.

3. AN ALGORITHM TO OPTIMIZE THE CALCULATION OF THE NUMBER OF STEPS IN THE RUNGE-KUTTA METHOD

The analysis given in Section 2 reveals that the Equations (3) behave chaotically depending on the number of steps. Previous research findings have tried to provide mathematical explanations for why such phenomena occur in nonlinear systems [35]. Our work does not intend to build upon these mathematical explanations. Instead, this work intends to create an algorithm that calculates the smallest number of steps necessary to solve the Equations (3), thereby preventing users/programs from encountering these chaotic behaviours. Along these lines, Hagebeuk and Kivits have presented an algorithm in the form of expansion of a very small parameter P_1^{-1} to overcome the problem of stiffness [35]. More recently, a new improved class of RK methods of the fourth order were applied with success in several sample problems [36-38]. Nevertheless, the problem of finding the smallest number of steps in which the solutions of RK methods are stable, numerically meaningful and efficiently calculated still remains a subject of great interest [39-42].

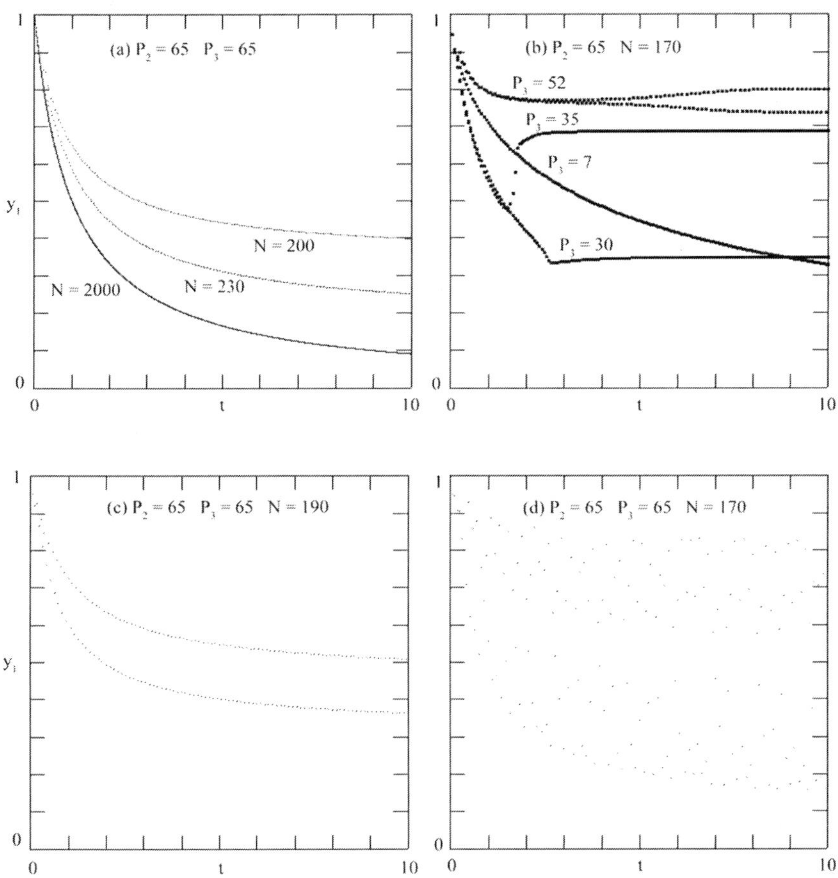

Figure 1. Solution y_1 of Equations (3) using the Runge-Kutta method.

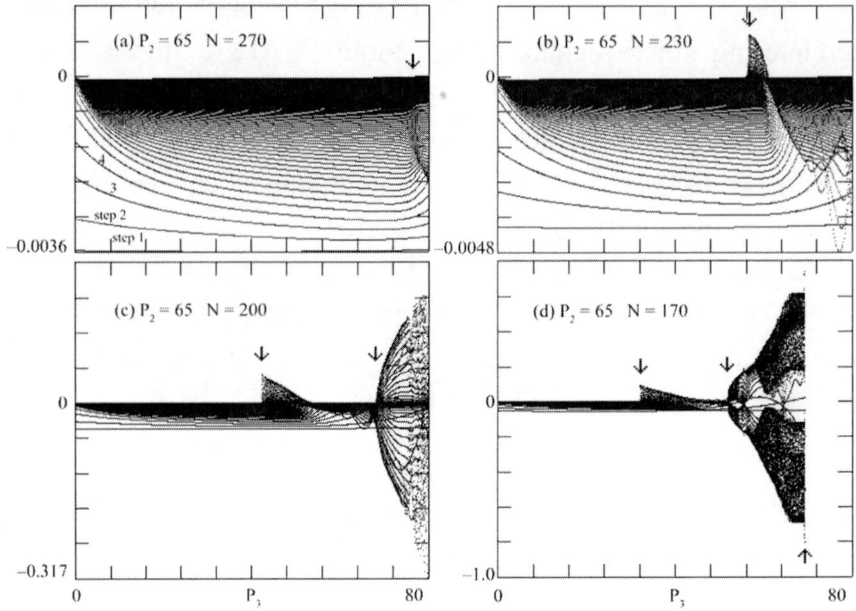

Figure 2. Plot of $y_1^{n+1} - y_1^n$ as function of $p_3 \in [0,80]$, with $p_2 = 65$, and (a) $N = 270$; (b) $N = 230$; (c) $N = 200$; and (d) $N = 170$.

3.1. How to Determine the Smallest Non-Critical Number of Steps?

The Runge-Kutta method numerically solves Equations (3) if the solution in a given interval becomes independent of the number of steps (i.e. the number of steps is large enough). However, as discussed in Section 2, we find that $y_1^{n+1} - y_1^n$ can detect fin-like, doubling and divergence (**Figure 2**), but cannot show incorrect solutions. This observation is disturbing because it reminds us of the need to perform additional tests to ensure that the number of steps is large enough to give reliable results for the numerical integration. Therefore, we propose an algorithm that executes this task automatically as the solution to find the smallest number of steps to finding a correct solution to Equations (3).

In Section 3, we observed that incorrect solutions may occur from the beginning of the calculations. For example, considering the same interval of integration, for $N = 200$, $p_1 = 1, p_2 = p_3 = 65$, which gives wrong results, y_1 in the first step is 0.954288... and for $N = 2000$, $p_1 = 1, p_2 = p_3 = 65$, which gives correct results, it is 0.952203... after 10 steps (the interval is the same in both calculations). This result shows that there is a difference from the beginning of the calculation. On the other hand for N = 2000, $p_1 = 1$, $p_2 = p_3 = 65$, the result of the calculation in the first step is 0.995014870675078 and for N = 20,000 after 10 steps the same level of numerical precision is achieved.

By examining these results, we conclude that the result of the application of the RK method using the interval of the first step and dividing it into 10 steps may reveal whether the number of steps will lead to incorrect results. So we will build the algorithm considering only the interval of the first step. We calculate the result of the application of the RK method with a single step in this interval and subtract the result of the calculation obtained by dividing this same interval into M steps. We write this algorithm as:

$$\Delta_i(M) = y_i(M) - y_i(1) \quad i = 1,2 \tag{5}$$

Let us consider the difference $\Delta_1(10)$ where $y_1(1)$ and $y_1(10)$ are calculated with one and 10 steps in the interval h, respectively. Then, since we are dealing with an expansion of the fourth order, we expect that to produce a correct solution this difference must be smaller than h^5. In case of incorrect solutions, the difference does not maintain proportional to h^5 and will increase.

Figure 3 shows $\Delta_1(10)$ as a function of the parameter $p_3 \in [0,80]$, with $p_2 = 65$ and the number steps (a) N = 270, (b) $N = 230$,

(c) $N = 200$, and (d) $N = 170$. In **Figure 3**(a) ($N = 270$) the arrow shows a sharp increase in $|\Delta_1(10)|$, at approximately $p_3 \approx 75$, which coincides with the region where the the results are wrong. In **Figure 3**(b) ($N = 230$) the arrow shows an abrupt increase in $|\Delta_1(10)|$, at about $p_3 \approx 56$, which coincides with the fin-like region. In**Figure 3**(c) ($N = 200$), the first arrow at approximately $p_3 \approx 21$, shows the beginning of a plateau-like region that extends to $p_3 \approx 42$, where the fin-like region indicated by a second arrow begins. In **Figure 3**(d) ($N = 170$), the first arrow shows the beginning of a plateau-like region extending to $p_3 \approx 42$ at about $p_3 \approx 9$, where the fin-like region indicated by a second arrow begins and the third arrow indicates divergence. The plateau-like region is not shown in the plot of $y_1^{n+1} - y_1^n$ (**Figure 2**).

Figure 4 shows the values of $\Delta_1(10)$ and $\Delta_2(10)$ as a function of the parameter $p_3 \in [0,80]$, with p2 = 65 and $N = 200$. The features of $\Delta_1(10)$ and $\Delta_2(10)$ are very similar, until the second arrow; above the second arrow there are differences in the structure of the results. Therefore, the best description of the error in the application of the Runge-Kutta method in the solution of Equations (3) may include both $\Delta_1(10)$ and $\Delta_2(10)$ features.

To include the behavior of $\Delta_1(10)$ and $\Delta_2(10)$, as suggested by the analysis of **Figure 4**, we chose the last term of equation 1b, because it contains the variables y_1 and y_2. Therefore we write the new algorithm $\Delta_{1,2}(M)$ as:

$$\begin{aligned}\Delta_{1,2}(M) = y_2(M)\left[y_2(M) + y_1(M)\right] \\ -y_2(1)\left[y_2(1) + y_1(1)\right]\end{aligned} \qquad (6)$$

Figure 5 shows $\Delta_{1,2}(10)$ as a function of $p_3 \in [0,80]$, with $p_2 = 65$ and (a) $N = 270$, (b) $N = 230$, (c) $N = 200$, and (d) $N = 170$.**Figure 5**(a) ($N = 270$) shows agreement with the exact solution until $p_3 \approx 75$, where a sharp increase begins. **Figure 5**(b) ($N = 230$) shows the exact solution region until $p_3 \approx 56$. **Figure 5**(c) ($N = 200$) shows the plateau-like region between two arrows. **Figure 5**(d) ($N = 170$) shows the plateau-like, fin-like and divergence regions successively.

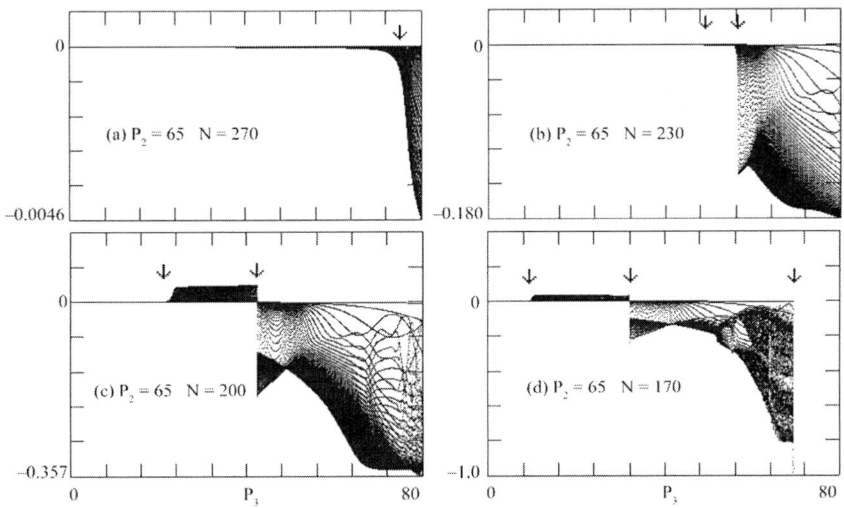

Figure 3. The difference $\Delta_1(10)$ shown as function of $p_3 \in [0,80]$, with $p_2 = 65$, and (a) $N = 270$; (b) $N = 230$; (c) $N = 200$; and (d) $N = 170$.

The results of $\Delta_1(10)$ and $\Delta_{1,2}(10)$ are similar for small values of the number of steps ($N = 170$ and $N = 200$). However when the number of steps increases closer to the number ideal of steps, we observed that $\Delta_{1,2}(10)$ reveals more details. For this reason, we adopted the algorithm $\Delta_{1,2}(10)$ to determine the smallest number of steps with which it is possible to reliably solve Equations (3) using the RK method.

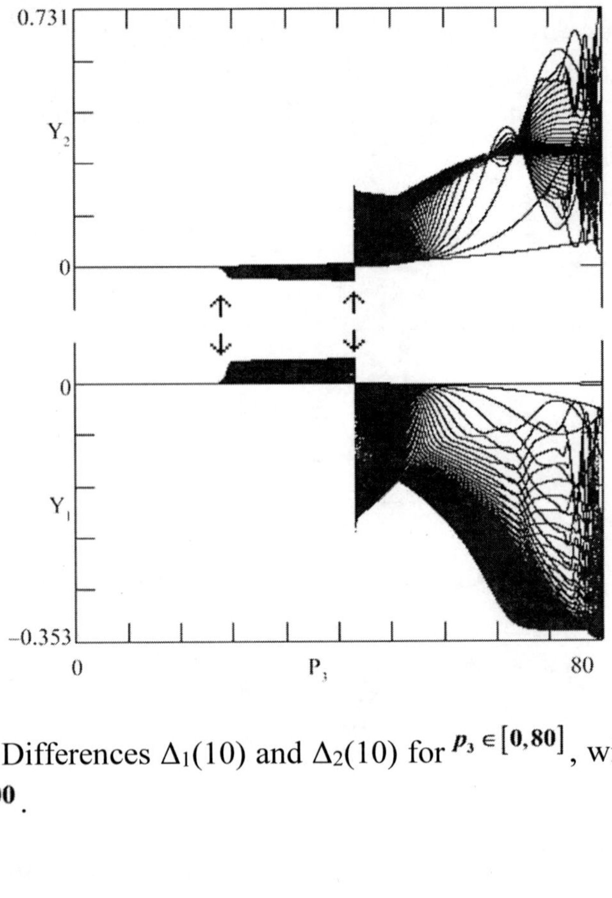

Figure 4. Differences Δ1(10) and Δ2(10) for $p_3 \in [0,80]$, with $p_2 = 65$, and $N = 200$.

Table 1 shows the number of steps and the values of y_1 at the end of the time interval determined using the algorithm $\Delta_{1,2}(10)$. For values of parameters $p_1 = 1,\ \ p_2 = p_3 = 65$, we see that the solution for $\Delta_{1,2}(10)$, is bifurcated for values equal to or greater than 0.004, but is convergent below 0.0039. We note that in the interval of values of delta between 0.0039 and E-06 the values of steps are equal and the values of y_1 match up to the third decimal place when compared to the more accurate results. The accuracy in the fifth decimal place is obtained in the interval of E-08 to E-10. For delta equal to E-08, the number of steps is 245 and for E-10 it is 340. For values of parameters $p_1 = 1, p_2 = 65, p_3 = 6$, we see that the solution is bifurcated for $\Delta_{1,2}(10)$ values equal to or greater than 0.02. Between 0.01 and 0.019, the solution is wrong. Between 0.004 and 0.001 accuracy is limited to the first decimal place, between 0.0001 and 0.00001 it extends to the third decimal place and for values less than 0.000001 to the fifth decimal place. For delta equal to E-06, the number of steps is 161 and 677 for E-10. In any case, we find that, even for very small values of delta, the magnitude of the number of steps is in an acceptable range of values.

3.2. The Algorithm

Taking into account the above considerations, we propose a 5-step algorithm to determine the smallest noncritical number of steps of the Runge-Kutta method applied to Equations (3), as follows:

- Define the starting number of steps (50 in the present case);
- Evaluate $\Delta_{1,2}(10)$ using Equation (6);
- Compare $\Delta_{1,2}(10)$ with the minimum acceptable error (E–10 in the present case);
- If $\Delta_{1,2}(10)$ does not match the minimum acceptable value, steps 2 and 3 are repeated with the number of steps increased by a constant value (50 in the present case);

- If $\Delta_{1,2}(10)$ matches the minimum acceptable value, the process is stopped.

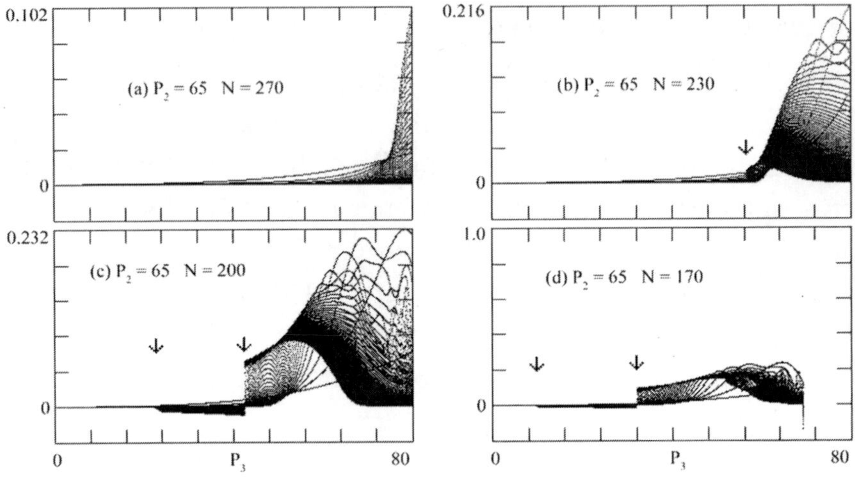

Figure 5. The $\Delta_{1,2}(10)$ difference shown as a function of $p_3 \in [0,80]$, with $p_2 = 65$, and (a) $N = 270$; (b) $N = 230$; (c) $N = 200$; and (d) $N = 170$

Applying the algorithm described above in the example presented in Section 3.1 for $p_1 = 1, p_2 = p_3 = 65$, we obtain $N = 340$ as the smallest non-critical number of steps. The last value calculated for y_1 in this case (N = 340) is 9.0907128174813E-02, while for $N = 2000$ it is 9.09071645004228E-02. Therefore the number of steps determined by our algorithm gives a solution that is correct and obtains a good degree of accuracy compared with that obtained using a larger number of steps.

Table 1. Values of the number of steps and the y_1 at the end of the time interval determined using the algorithm $\Delta_{1,2}(10)$ with values equal or smaller that those shown in the first column. Values of N^a and y_1^a were determined for $p_1 = 1, p_2 = p_3 = 65$ and of N^b and y_1^b were determined for $p_1 = 1, p_2 = 65, p_3 = 6$.

$\Delta_{1,2}(10)$	N^a	y_1^a	N^b	y_1^b
1E-01	131	0.388058	53	0.841878
2E-02	131	0.388098	151	0.858205
1.9E-02	131	0.388098	153	0.434904
1E-02	131	0.388058	151	0.331284
4E-03	164	0.753236	153	0.348315
3.9E-03	241	0.090869	153	0.348315
1E-03	241	0.090869	153	0.348315
1E-04	241	0.090869	155	0.350197
1E-05	241	0.090869	158	0.350395
1E-06	241	0.090869	161	0.350405
1E-07	242	0.090897	168	0.350407
1E-08	245	0.090909	225	0.350406
1E-09	266	0.090906	366	0.350406
1E-10	340	0.090907	677	0.350405

To test our algorithm, we implemented a function using the programming language Basic. Part of the code is shown on **Table 2**. Lines 1 to 14 represent the first step of the algorithm, lines 15 to 33 the second step, lines 34 and 35 the third step, 36 to 40 the fourth step and 41 to 46 the last step. We performed a series of tests changing the parameters of the equations and checking the solution for Adirovitch's model. To the best of our knowledge, our algorithm has produced only correct solutions.

It is worth noting that we also successfully used the same algorithm to calculate the minimum number of steps to find the correct solution to more complex equations that describes the kinetics of atomic hydrogen [43]. We are aware of other solutions for problems in the Runge-Kutta method [38,39,42,44]. Nevertheless, some of them are not open source (e.g. algorithm presented in MathLab) and others cannot be applied in our previously developed programs and algorithms without extensive changes.

We are currently developing software to simulate and fit the experimental results of decay to time and thermoluminescence using the present algorithm for determining the number of steps, optimizing the parameters using the grid method described in previous work [27]. Preliminary results show perfect integration among these, with safety and manageability.

4. CONCLUSIONS

The calculation of solutions for differential equations can be complex and time consuming, requiring large amounts of computational resources (processing time, memory, etc). Thus, it is essential to optimize this process in order to get faster results. In this paper we were interested in using the Runge-Kutta method to calculate the correct solution for the Adirovitch model that uses coupled rate differential equations to describe the kinetic electron trapping process in a shallow defect state and its subsequent thermalor photo-stimulated promotion to a conduction band followed by recombination in another defect.

A series of experiments were carried out, demonstrating that using the Runge-Kutta Fourth Order method to calculate the solution for the Adirovitch model's differential equations may lead to incorrect results when the numbers of steps to calculate the solution is not "large enough". Thus, a numerical analysis was conducted. As a result we develop an algorithm that automatically determines the smallest non-critical number of steps that optimize the calculation of correct solutions of differential equations that follow the same behaviour found in the Adirovitch model. The algorithm is composed of 5 steps: 1) Define the starting number of steps; 2) Evaluate $\Delta_{1,2}(10)$ using Equation (6); 3) Compare $\Delta_{1,2}(10)$ with the minimum acceptable error; 4) If $\Delta_{1,2}(10)$ does not match the minimum acceptable value, steps 2 and 3 are repeated with the number of steps increased by a constant value (50 in the present case); 5) If $\Delta_{1,2}(10)$ matches the minimum acceptable value, the process is stopped. This algorithm was implemented and tested with different values and parameters. We also successfully used the same algorithm to optimize the calculation of more complex equations that describe the kinetics of atomic hydrogen [43].

In future work we will extend this algorithm to support fast calculation of different models that describe physical behaviors using differential equations. We will also use the results of this work to create software that works as a platform to conduct experiments that simulate the behaviors of different models. It will enable users to change the values of parameters in the equations and rapidly check the implications in the behavior of each model.

Table 2. Part of the code of the proposed algorithm implemented in Basic.

```
Private Sub FindNRungeKutta()
FlagCancel = False    [indicates if the solution diverges]
NRFIM = 50                    [initial steps]
LeituradeParametros [read the equation parameters]
For ip = 1 To IXE - 1              [the calculation is done on all the intervals
                                                  between the data being simulated]
    ContinuaFNRKAn1101:
    If NRFIM > 1000000 Then
        msg = "The degree of Stiffness does not allow the calculation of the solution"
        MsgBox msg, 48, " NOTE"
        FlagCancel = True [solution diverges]
    End
    H1 = (XEXP(ip + 1) - XEXP(ip)) / NRFIM
        [Calculate steps of the integration for interval ip e ip+1]

    For ip2 = 1 To 2
        [for ip2=1 calculate the second parameter of equation (6) using step H1    (line 13).
        For ip2=2 calculate the first parameter of equation (6) with M=10 using the redefined H1 (line 31)]
        Z1 = 1#      [initial condition for Y1]
        Z2 = 0          [initial condition for Y2]
        Z0 = XEXP(ip)
            [initial conditions for the Runge Kutta algorithm]
        Sub2700
```

```
            [function that calculates diferential equations using Runge Kutta and returns YCALC(ip2)]
        If FlagCancel = True Then                   [equation diverges => too few steps]
            NRFIM = NRFIM + 50          [increase number of steps]
            FlagCancel = False              [restart flag]
            GoTo ContinuaFNRKAn1101    [restart algorithm]
        End If
        H1 = H1 * 0.1
        NU = 10 * NU
    Next ip2

    If Abs(1# - YCALC(1) / YCALC(2)) > 0.0001 Then
        [compare the terms of the algoritm  Δ_{1,2}(M) ]

        NRFIM = NRFIM + 50
        [increase the number of steps for integration]
        GoTo ContinuaFNRKAn1101    [return the results for  Δ_{1,2}(M) ]
    Else

        NFIM = NRFIM [accepts calculated value]
        Text11.Text = Format(NFIM, "#########") [print it]
        FlagCancel = False
    End If
Next ip
End Sub
```

ACKNOWLEDGEMENTS

This work was supported in part by grants from FINEP, CNPq and FAPESP.

REFERENCES

1. E. I. Adirovitch, "The Formula of Becquerel and the Elementary Decay Law of Luminescence of Crystal Phosphors," Journal of Physique Archives (Paris), Vol. 17, No. 8-9, 1956, pp. 705-707.

2. S. W. S. McKeever and R. Chen, "Luminescence Models," Radiation Measurements, Vol. 27, No. 5-6, 1997, pp. 625-661.

3. C. M. Sunta, W. E. F. Ayta, R. N. Kulkarni, T. M. Piters and S. Watanabe, "Theoretical Models of Thermoluminescence and Their Relevance in Experimental Work," Radiation Protection Dosimetry, Vol. 84, No. 1-4, 1999, pp. 25-28.

4. S. Isotani, W. W. Furtado, R. Antonini and O. L. Dias, "Line-Shape and Thermal Kinetics Analysis of the Fe^{2+}- Band in Brazilian Green Beryl," American Mineralogist, Vol. 74, No. 1, 1989, pp. 432-438.

5. R. Antonini, S. Isotani, W. W. Furtado, W. M. Pontuschka and S. R. Rabbani, "Study of the Decay Kinetics of Irradiation Induced Green Color in Brazilian Spodumene," Proceedings of the Brazilian Academy of Sciences, Vol. 62, No. 1, 1990, pp. 39-43.

6. A. S. Ito and S. Isotani, "Heating Effects on the Optical Absorption Spectra of Irradiated, Natural Spodumene," Radiation Effects and Defects in Solids: Incorporating Plasma

Science and Plasma Technology, Vol. 116, No. 4, 1991, pp. 307-314.

7. S. Isotani, A. T. Fujii, R. Antonini, W. M. Pontuschka, S. R. Rabbani and W. W. Furtado, "Luminescence Study of Spodumene," Proceedings of the Brazilian Academy of Sciences, Vol. 62, No. 2, 1990, pp. 107-113.

8. S. Isotani, W. W. Furtado, R. Antonini, A. R. Blak, W. M. Pontuschka, T. Tomé and S. R. Rabbani, "Decay Kinetics Study of Atomic Hydrogen in a-Si:(H,O,N) and Natural Beryl," Physical Review B, Vol. 42, No. 10, 1990, pp. 5966-5972.

9. M. B. Camargo, W. M. Pontuschka and S. Isotani, "Electron Paramagnetic Resonance of Atomic Hydrogen Centers in Rubellite," Proceedings of the Brazilian Academy of Sciences, Vol. 59, No. 1, 1987, pp. 293-298.

10. A. R. Blak, W. M. Pontuschka and S. Isotani, "Electron Paramagnetic Resonance of Hydrogen Centers in Natural Beryl," Proceedings of the Brazilian Academy of Sciences, Vol. 60, No. 1, 1988, pp. 9-12.

11. S. Sciortino, "Growth, Characterization and Properties of CVD Diamond Films for Applications as Radiation Detectors," La Rivista del Nuovo Cimento, Vol. 22, No. 10, 1999, pp. 1-89.

12. T. Budinas, A. Jankauskas, P. Mackup and A. Smilga, "Charge Transport Peculiarities in Thin-Layer Photodielectric System $As_2Se_3Tl_x$-Sb_2S_3," Thin Solid Films, Vol. 36, No. 2, 1976, pp. 277-280.

13. A. E. do Nascimento, P. Trzesniak, M. E. G. Valerio and J. F. de Lima, "On the Error in the Activation Energy Obtained by the Initial Raise Method for Thermally Stimulated Processes in Dielectrics," Radiation Effects and Defects in Solids:

Incorporating Plasma Science and Plasma Technology, Vol. 134, No. 1-4, 1995, pp. 147-152.

14. A. Opanowicz, “Interactive Kinetics in Thermally Stimulated Luminescence in Insulators and Semiconductors,” Physica B: Condensed Matter, Vol. 271, No. 1-4, 1999, pp. 273-283.

15. L. Brizhik, F. Musumeci, A. Scordino and A. Triglia, “The Soliton Mechanism of the Delayed Luminescence of Biological Systems,” EPL (Europhysics Letters), Vol. 52, No. 2, 2000, pp. 238-244.

16. R. Biswas and B. C. Pan, “Defect Kinetics in New Model of Metastability in a-Si:H,” Journal of Non-Crystalline Solids, Vol. 299-302, Part 1, 2002, pp. 507-510.

17. A. D. Drozdov, J. De C. Christiansen, R. Klitkou and C. G. Potarniche, “Effect of Annealing on Viscoplasticity of Polymer Blends: Experiments and Modeling,” Computational Materials Science, Vol. 50, No. 1, 2010, pp. 59-64.

18. W. W. Furtado, T. Tomé, S. Isotani, R. Antonini, A. R. Blak, W. M. Pontuschka and S. R. Rabbani, “Numerical Integration Method Applied to the Study of Atomic Hydrogen in Aluminoborate Glass,” Proceedings of the Brazilian Academy of Sciences, Vol. 61, No. 4, 1989, pp. 397-403.

19. A. Mizukami, S. Isotani, S. R. Rabbani and W. M. Pontuschka, “Approximate Solution for Kinetic Differential Equations,” Il Nuovo Cimento D, Vol. 15, No. 4, 1993, pp. 637-645.

20. P. Kelly, M. J. Laubitz and P. Braunlich, “Exact Solutions of the Kinetic Equations Governing Thermally Stimulated Luminescence and Conductivity,” Physical Review B, Vol. 4, No. 6, 1971, pp. 1960-1968.

21. J. C. Butcher, "Runge-Kutta Methods in Modern Computation, Part I: Fundamental Concepts," Computers in Physics, Vol. 8, No. 4, 1994, pp. 411-415.

22. P. Rentrop, "Partitioned Runge-Kutta Methods with Stiffness Detection and Stepsize Control," Numerische Mathematik, Vol. 47, No. 4, 1985, pp. 545-564.doi:10.1007/BF01389456

23. A. Brunetti, "A Fast and Precise Genetic Algorithm for a Non-Linear Fitting Problem," Computer Physics Communications, Vol. 124, No. 2-3, 2000, pp. 204-211.doi:10.1016/S0010-4655(99)00454-3

24. F. Khorasheh, A. M. Ahmadi and A. Gerayeli, "Application of Direct Search Optimization for Pharmacokinetic Parameter Estimation," Journal of Pharmacy and Pharmaceutical Science, Vol. 2, No. 3, 1999, pp. 92-98.

25. P. R. Bevington and D. K. Robinson, "Data Reduction and Error Analysis for the Physical Science," 2nd Edition, McGraw-Hill, New York, 1992.

26. S. Isotani and A. T. Fujii, "A Grid Procedure Applied to the Determination of Parameters of a Kinetic Process," Computer Physics Communications, Vol. 151, No. 1, 2003, pp. 1-7.

27. W. H. Press, B. P. Flannery, S. A. Teukolsky and W. T. Vetterling, "Numerical Recipes in C++: The Art of Scientific Computation," Cambridge University Press, New York, 1988.

28. D. W. Zingg and T. T. Chisholm, "Runge-Kutta Methods for Linear Ordinary Differential Equations," Applied Numerical Mathematics, Vol. 31, No. 2, 1999, pp. 227-238.

29. R. I. Mclachlan, Y. Sun and P. S. P. Tse, "Linear Stability of Partitioned Runge-Kutta Methods," SIAM Journal on Numerical Analysis, Vol. 49, No. 1, 2011, pp. 232-263.

30. G. Dimarco and L. Pareschi, "Exponential Runge-Kutta Methods for Stiff Kinetic Equations," SIAM Journal on Numerical Analysis, Vol. 49, No. 5, 2011, pp. 2057-2077.

31. J. Stoer and R. Bulirsch, "Introduction to Numerical Analysis," Springer Verlag, New York, 1980.

32. C. W. Cear, "Numerical Initial Value Problems in Ordinary Differential Equations," Prentice Hall, Upper Saddle River, 1971.

33. K. Gurski and S. O'Sullivan, "A Stability Study of a New Explicit Numerical Scheme for a System of Differential Equations with a Large Skew-Symmetric Component," SIAM Journal on Numerical Analysis, Vol. 49, No. 1, 2011, pp. 368-386.

34. M. J. Feigembaum, "Universal Behavior in Nonlinear Systems," Physica D: Nonlinear Phenomena, Vol. 7, No. 1-3, 1983, pp. 16-39.

35. H. J. L. Hagebeuk and P. Kivits, "Determination of Trapping Parameters from the Conventional Model for Thermally Stimulated Luminescence and Conductivity," Physica B+C, Vol. 83, No. 3, 1976, pp. 289-294.

36. A. E. Aboanber, "Stability of Generalized Runge-Kutta Methods for Stiff Kinetics Coupled Differential Equations," Journal of Physics A: Mathematical and General, Vol. 39, No. 8, 2006, pp. 1859-1876.

37. X. J. Wang and S. Q. Gan, "A Runge-Kutta Type Scheme for Nonlinear Stochastic Partial Differential Equations with Multiplicative Trace Class Noise," Numerical Algorithms, 2012. (To Be Published).

38. H. Yuan, J. Zhao and Y. Xu, "Some Stability and Convergence of Additive Runge-Kutta Methods for Delay Differential Equations with Many Delays," Journal of Applied Mathematics, Vol. 2012, No. 1, 2012, pp. 1-18.

39. J. S. C. Prentice, "Stepsize Selection in Explicit RungeKutta Methods for Moderately Stiff Problems," Applied Mathematics, Vol. 2, No. 6, 2011, pp. 711-717.

40. M. N. Spijker, "Stepsize Conditions for General Monotonicity in Numerical Initial Value Problems," SIAM Journal on Numerical Analysis, Vol. 45, No. 3, 2007, pp. 1226-1245.

41. S. Gottlieb, D. I. Ketcheson and C.-W. Shu, "Strong Stability Preservation Runge-Kutta and Multistep Time Discretizations," World Scientific Publishing Co., Singapore, 2011.

42. A. Valinejad and S. M. Hosseini, "A Stepsize Control Algorithms for SDEs with Small Noise Based on Stochastic Runge-Kutta Maruyama Methods," Numerical Algorithms, Vol. 61, No. 3, 2012, pp. 479-498. W. M. Pontuschka and S. Isotani, "A Model for the Stabilization of Atomic Hydrogen Centers in Borate Glasses," Physica B: Condensed Matter, Vol. 404, No. 21, 2009, pp. 4312-4315.

43. D. I. Ketcheson and A. J. Ahmadia, "Optimal RungeKutta Stability Regions," 2012. http://arxiv.org/pdf/1201.3035v2.pdf

CHAPTER 3

Parabolic Partial Differential Equations as Inverse Moments Problem

Maria B. Pintarelli[1,2]

[1] Grupo de Aplicaciones Matematicas y Estadisticas de la Facultad de Ingenieria (GAMEFI), Universidad Nacional de La Plata (UNLP), La Plata, Argentina

[2] Departamento de Matematica, Facultad de Ciencias Exactas, Universidad Nacional de La Plata (UNLP), La Plata, Argentina

ABSTRACT

We considerer parabolic partial differential equations $w_t - (w_x)_x = r(x,t)$ under the conditions $w_x(a_1,t) = k_1(t)$ $w_x(b_1,t) = k_2(t)$ $w(x,a_2) = h_1(t)$ $w(x,b_2) = h_2(t)$ on a region $E = (a_1,b_1) \times (a_2,b_2)$. We will see that we can write the equation in partial derivatives as an Fredholm integral equation of first kind and will solve this latter with the techniques of inverse moments problem. We will find an approximated solution and bounds for the error of the estimated solution using the techniques on moments problem. Also we consider the one- dimensional one-phase inverse Stefan problem.

KEYWORDS

Parabolic PDEs, Freholm Integral Equations, Generalized Moment Problem

1. INTRODUCTION

We considerer parabolic partial differential equation of the form:

$$w_t - (w_x)_x = r(x,t) \tag{1}$$

where the unknown function $w(x,t)$ is defined in $E=(a_1,b_1)\times(a_2,b_2)$. $r(x,t)$ is known function. We consider conditions

$$w_x(a_1,t)=k_1(t) \qquad w_x(b_1,t)=k_2(t)$$

$$w(x,a_2)=h_1(t) \qquad w(x,b_2)=h_2(t).$$

This problem was studied under conditions of Cauchy in [1] .

Parabolic differential equations are commonly used in the fields of engineering and science for simulating physical processes. These equations describe various processes in viscous fluid flow, filtration of liquids, gas dynamics, heat conduction, elasticity, biological species, chemical reactions, environmental pollution, etc. [2] [3] .

In a variety of cases, approximations are used to convert parabolic PDEs to ordinary differential equations or even to algebraic equations. The existence and uniqueness properties of this problem are presented in literature. Several numerical methods have been proposed for the solution of this problem [4] -[6] .

Next section is devoted to showing how the differential equation (1) is transformed into integral equation of first kind that can be seen as generalized moments problem. In Section 3 there we present a theorem that guarantees under certain conditions the stability and convergence of the finite generalized moment problem. In Section 4 we exemplify the general method by applying it to some parabolic PDEs of the form (1). Finally in Section 5 the method is applied to solve the one-dimensional one-phase inverse Stefan problem.

The Stefan problem consists of finding w y s such that

$$\frac{\partial w}{\partial t}=\frac{\partial^2 w}{\partial x^2} \quad \text{in } \{(x,t);0<x<s(t);t>0\}$$

$$-\frac{\partial w}{\partial x}(0,t)=f(t) \qquad t>0$$

$$w(s(t),t)=0 \qquad t>0$$

$$\frac{ds}{dt}=-\frac{\partial w}{\partial x}(s(t),t) \qquad t>0$$

$$w(x,0)=w_0(x) \qquad x\geq 0$$

$$s(0)=a.$$

The classical Stefan problem is a nonlinear initial value problem with a moving boundary whose position is unknown a priori and it must be determined as part of the solution. The differential equations of parabolic type governing heat diffusion with phase change are an important class of Stefan problems.

The direct Stefan problem requires determining both the temperature and the moving boundary interface when the initial and boundary conditions, and the thermal properties of the heat conducting body are known. Conversely, inverse Stefan problems require determining the initial and/or boundary conditions, and/or thermal properties from additional information which may involve the partial knowledge or measurement of the moving boundary interface position, its velocity in a normal direction, or the temperature at selected interior thermo- couples of the domain.

In this paper we solve the inverse Stefan problem: find $f(t)$ with $s(t)$ known such that the above con- ditions are met.

The d-dimensional generalized moment problem [7] -[10] can be posed as follows: find a function u on a domain $\Omega\subset\mathbf{R}^d$ satisfying the sequence of equations

$$\int_{\Omega} u(x)g_n(x)\mathrm{d}x = \mu_n \qquad n\in\mathbf{N} \tag{2}$$

where (g_n) is a given sequence of functions lying in $\mathbf{L}^2(\Omega)$ linearly independent.

Many inverse problems can be formulated as an integral equation of the first kind, namely,

$$\int_a^b K(x,y)u(y)\mathrm{d}y = f(x) \quad x\in(a,b).$$

$K(x,y)$ and $f(x)$ are given functions and $u(y)$ is a solution to be determined; is a result of experimental measurements and hence is given only at finite set of points. It follows that the above integral equation is equivalent to the following moment problem

$$\int_a^b K(x_n, y)u(y)\mathrm{d}y = f(x_n) \qquad n = 1,2,\cdots.$$

Also we considerer the multidimensional moment problems

$$\int_\Omega K(x_n, y)u(y)\mathrm{d}y = f(x_n) \qquad n = 1,2,\cdots,\Omega \subset \mathbf{R}^d.$$

Moment problems are usually ill-posed. There are various methods of constructing regularized solutions, that is, a approximate solution stable with respect to the given data. One of them is the method of truncated expansion.

The method of truncated expansion consists in approximating (2) by finite moment problems

$$\int_\Omega u(x)g_i(x)\mathrm{d}x = \mu_i \qquad i = 1,2,\cdots,n. \tag{3}$$

Solved in the subspace $\langle g_1, g_2, \cdots, g_n \rangle$ generated by $g_1, g_2, \cdots, g_n$ (3) is stable. Considering the case where the data $\mu = (\mu_1, \mu_2, \cdots, \mu_n)$ are inexact, we apply some convergence theorems and error estimates for the regularized solutions [9] [11].

2. PARABOLIC PARTIAL DIFFERENTIAL EQUATIONS AS INTEGRAL EQUATIONS OF FIRST KIND

Let $F(w(x,t)) = r(x,t)$ be a partial differential equations such as (1). The solution $w(x,t)$ is defined on the region $E = (a_1, b_1) \times (a_2, b_2)$ and verifies conditiones on the boundary $C = \partial E$:

$$w_x(a_1,t)=k_1(t) \qquad w_x(b_1,t)=k_2(t)$$

$$w(x,a_2)=h_1(t) \qquad w(x,b_2)=h_2(t).$$

We apply the technique used in [1] . Let $F^*=(F_1(w),F_2(w))$ be a vectorial field such that w verifies $div(F^*)=h^*(w)$ with h^* a known function and, reciprocally, if w verifies $div(F^*)=h^*(w)$ then

$$F(w(x,t))=r(x,t).$$

Specifically in this case $F(w(x,t))=w_t-(w_x)_x$ and we take

$$F^*=(F_1(w),F_2(w))=(w_x,-w) \tag{4}$$

$$h^*(w)=-r(x,t). \tag{5}$$

Let $u(x,t,\tau,\xi)$ be the auxiliary function

$$u(x,t,\tau,\xi)=\mathrm{e}^{-t\xi}\cos\left(\frac{x\pi}{b_1}\tau\right). \tag{6}$$

Since

$$udiv(F^*)=uh^*(w)$$

we have

$$\iint_E udiv(F^*)\mathrm{d}A=\iint_E uh^*(w)\mathrm{d}A.$$

Moreover, as

$$udiv\left(F^{*}\right)=div\left(uF^{*}\right)-F^{*}\cdot\nabla u$$

And

$$\iint_{E}udiv\left(F^{*}\right)\mathrm{d}A=\underbrace{\iint_{E}div\left(uF^{*}\right)\mathrm{d}A}_{=\int_{C}\left(uF^{*}\right)nds}-\iint_{E}F^{*}\cdot\nabla u\mathrm{d}A$$

we obtain

$$\iint_{E}uh^{*}\left(w\right)\mathrm{d}A=\int_{C}\left(uF^{*}\right)\cdot nds-\iint_{E}F^{*}\cdot\nabla u\mathrm{d}A \tag{7}$$

where $\nabla u=\left(u_{\tau},u_{\xi}\right)$.

We consider the integral

$$\iint_{E}F^{*}\cdot\nabla u\mathrm{d}A=\iint_{E}F_{1}u_{\tau}+F_{2}u_{\xi}\mathrm{d}A. \tag{8}$$

Integrating by parts:

$$\begin{aligned}\iint_{E}F_{1}u_{\tau}\mathrm{d}A=\int_{a_2}^{b_2}\int_{a_1}^{b_1}F_{1}u_{\tau}\mathrm{d}\tau\mathrm{d}\xi&=\int_{a_2}^{b_2}\left(w\left(b_1,\xi\right)u_{\tau}\left(x,t,b_1,\xi\right)-w\left(a_1,\xi\right)u_{\tau}\left(x,t,a_1,\xi\right)\right)\mathrm{d}\xi\\&-\int_{a_2}^{b_2}\int_{a_1}^{b_1}w\left(\tau,\xi\right)u_{\tau\tau}\mathrm{d}\tau\mathrm{d}\xi.\end{aligned} \tag{9}$$

Note that in (9) if x is a natural number then

$$u_{\tau}\left(x,t,b_1,\xi\right)=-\mathrm{e}^{-t\xi}\frac{x\pi}{b_1}\sin\left(\frac{x\pi}{b_1}b_1\right)=0 \tag{10}$$

and if $a_1=0$ then

$$u_\tau\left(x,t,a_1,\xi\right)=-\mathrm{e}^{-t\xi}\frac{x\pi}{b_1}\sin\left(\frac{x\pi}{b_1}a_1\right)=0. \tag{11}$$

Thus if $x\in N$ and $a_1=0$:

$$\iint_E F^*\cdot\nabla u dA=-\iint_E\left(wu_{\tau\tau}-wu_\xi\right)\mathrm{d}A=-\int_{a_2}^{b_2}\int_{a_1}^{b_1}wu\left(\left(\frac{x\pi}{b_1}\right)^2+t\right)\mathrm{d}\tau\mathrm{d}\xi. \tag{12}$$

Also if we write $C=\partial E=C_1\cup C_2\cup C_3\cup C_4$ (see Figure 1) then

$$\int_C\left(uF^*\right)\cdot n\mathrm{d}s=\int_{C_1}\left(uF^*\right)\cdot n\mathrm{d}s+\int_{C_2}\left(uF^*\right)\cdot n\mathrm{d}s+\int_{C_3}\left(uF^*\right)\cdot n\mathrm{d}s+\int_{C_4}\left(uF^*\right)\cdot n\mathrm{d}s$$

$$\int_{C_1}\left(uF^*\right)\cdot n\mathrm{d}s=\int_{a_1}^{b_1}u\left(x,t,\tau,a_2\right)\underbrace{w\left(\tau,a_2\right)}_{h_1(\tau)}\mathrm{d}\tau$$

$$\int_{C_3}\left(uF^*\right)\cdot n\mathrm{d}s=-\int_{a_1}^{b_1}u\left(x,t,\tau,b_2\right)\underbrace{w\left(\tau,b_2\right)}_{h_2(\tau)}\mathrm{d}\tau$$

We write

$$G(x,t)=\int_{C_1}\left(uF^*\right)\cdot n\mathrm{d}s+\int_{C_2}\left(uF^*\right)\cdot n\mathrm{d}s+\int_{C_3}\left(uF^*\right)\cdot n\mathrm{d}s+\int_{C_4}\left(uF^*\right)\cdot n\mathrm{d}s$$

$$R(x,t)=\iint_E r(\tau,\xi)u(x,t,\tau,\xi)\mathrm{d}A$$

finally, if $x\in N$ y $a_1=0$ we get:

$$\int_{a_2}^{b_2}\int_{a_1}^{b_1}w(\tau,\xi)u(x,t,\tau,\xi)\mathrm{d}\tau\mathrm{d}\xi=\frac{G(x,t)-R(x,t)}{\left(\left(\frac{x\pi}{b_1}\right)^2+t\right)}=\varphi(x,t). \tag{13}$$

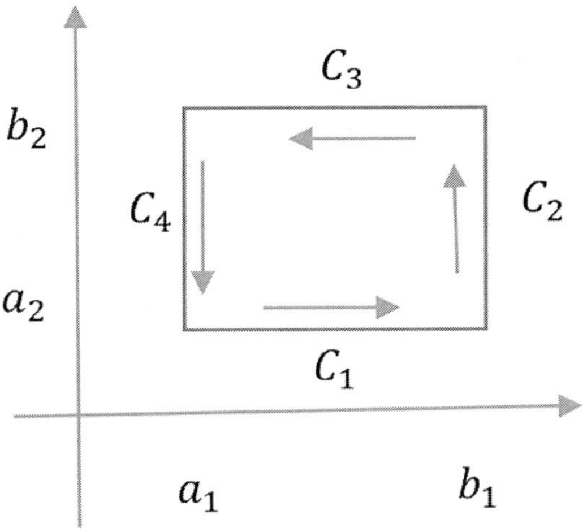

Figure 1. Domain E and contour $\partial E = C$.

If $E = \{(\tau,\xi)/\tau \geq 0; a_2 \leq \xi \leq b_2\}$ then you can take

$$u(x,t,\tau,\xi) = \mathrm{e}^{-t\xi}\cos(x\tau) \tag{14}$$

and we must have when $\tau \to \infty$.

3. SOLUTION OF GENERALIZED MOMENT PROBLEMS

Equation (13) is of the form:

$$\int_{a_1}^{b_1}\int_{a_2}^{b_2} w(\tau,\xi)K(x,t,\tau,\xi)\mathrm{d}\tau\mathrm{d}\xi = \varphi(x,t).$$

We assign natural values to x and t: $x=i, i=0,1,\cdots,m$ and $t=j, j=0,1,\cdots,n$ and we consider the corre- sponding generalized finite moment problem bi-dimensional [12] [13]

$$\int_{a_1}^{b_1}\int_{a_2}^{b_2} w(\tau,\xi)K_{ij}(\tau,\xi)\mathrm{d}\tau\mathrm{d}\xi=\varphi(i,j)=\mu_{ij} \quad i=0,1,\cdots,m \quad j=0,1,\cdots,n.$$

To obtain a numerical approximation of the solution $w(\tau,\xi)$ the truncated expansion method is applied [9] [11] .

We considerer the basis $\{\phi_i(\tau,\xi)\}_{i=0}^{\infty}$ obtained from the sequence $\{K_{ij}(\tau,\xi)\}$ with $i=0,1,\cdots,m;\ j=0,1,\cdots,n$ by Gram-Schmidt method and addition of the necessary functions in order to have an orthonormal basis.

To facilitate the calculations we write $\{K_i(\tau,\xi)\}_{i=0}^{r}$ and $\{\mu_i\}_{i=0}^{r}$ with $r=m\cdot n.$

We then approximate the solution $w(\tau,\xi)$ with

$$p_r(\tau,\xi)=\sum_{i=0}^{r}\lambda_i\phi_i(\tau,\xi)$$

and

$$\lambda_i=\sum_{j=0}^{i}C_{ij}\mu_j \qquad i=0,1,\cdots,r$$

where the coefficients C_{ij} verifies

$$C_{ij}=\left(\sum_{k=j}^{i-1}(-1)\frac{\langle K_i(\tau,\xi)\,|\,\phi_k(\tau,\xi)\rangle}{\|\phi_k(\tau,\xi)\|^2}C_{kj}\right)\cdot\|\phi_i(\tau,\xi)\|^{-1} \quad 1<i\le r;\ 1\le j<i \tag{15}$$

$$C_{ii} = \left\|\phi_i(\tau,\xi)\right\|^{-1} \qquad i = 0,1,\cdots,r. \tag{16}$$

The proof of the following theorem is in [14] .

Theorem 1. Let $\{\mu_m\}_{m=0}^{r}$ be a set of real numbers and let ε and E be two positive numbers such that

$$\sum_{m=0}^{r}\left|\int_{a_2}^{b_2}\int_{a_1}^{b_1}K_m(\tau,\xi)\,w(\tau,\xi)\,\mathrm{d}\tau\mathrm{d}\xi - \mu_m\right|^2 \le \varepsilon^2 \tag{17}$$

$$\int_{a_2}^{b_2}\int_{a_1}^{b_1}\left[(b_1-a_1)^2 w_\tau^2 + (b_2-a_2)^2 w_\xi^2\right]\mathrm{d}\tau\mathrm{d}\xi \le E^2 \tag{18}$$

then

$$\int_{a_2}^{b_2}\int_{a_1}^{b_1}\left|w(\tau,\xi)\right|^2 \mathrm{d}\tau\mathrm{d}\xi \le \min_m\left\{\left\|CC^{\mathrm{T}}\right\|\varepsilon^2 + \frac{E^2}{8(m+1)^2}; m = 0,1,\cdots,r\right\} \tag{19}$$

where C is the triangular matriz with elements C_{ij} $(1 < i \le r; 1 \le j < i)$.

And

$$\int_{a_2}^{b_2}\int_{a_1}^{b_1}\left|p_r(\tau,\xi) - w(\tau,\xi)\right|^2 \mathrm{d}\tau\mathrm{d}\xi \le \left\|CC^{\mathrm{T}}\right\|\varepsilon^2 + \frac{E^2}{8(r+1)^2}. \tag{20}$$

If $E = (a_1,\infty)\times(a_2,\infty)$, then (18) it is replaced by

$$\int_{a_2}^{\infty}\int_{a_1}^{\infty}\left[\tau w_\tau^2 + \xi w_\xi^2\right]\mathrm{Exp}[\tau+\xi]\,\mathrm{d}\tau\mathrm{d}\xi \le E^2$$

and we must have

$$\tau^m w(\tau,\xi) \to 0 \quad \text{if } \tau \to \infty \quad \text{for all } m \in N$$

and

$$\xi^n w(\tau,\xi) \to 0 \quad \text{if } \xi \to \infty \quad \text{for all } n \in N.$$

4. NUMERICAL EXAMPLES

4.1. Example 1

Let considerer the equation

$$w_t - w_{xx} = 0$$

in the domain $E = (0,2)\times(0,2)$

and boundary condition on ∂E

given by

$$w_x(0,t) = \mathrm{e}^{-t} \qquad w_x(2,t) = \cos(2)\mathrm{e}^{-t}$$

$$w(x,0) = \sin(x) \qquad w(x,2) = \sin(x)\mathrm{e}^{-2}.$$

The exact solution is $w(x,t) = \sin(x)\mathrm{e}^{-t}$.

In Figure 2 the approximate numerical solution and the exact one are compared.Were taken $r = 9$ moments.

The accuracy is, in this case $\int_0^2\int_0^2 |p_9(x,t) - w(x,t)|^2 \,\mathrm{d}x\mathrm{d}t = 0.0361472.$

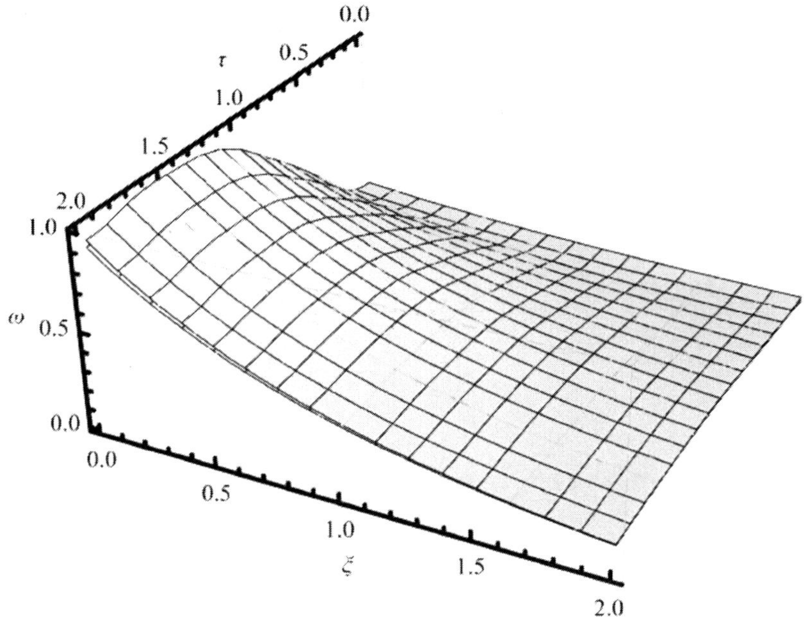

Figure 2. $w(x,t)=\sin(x)e^{-t}$ and $p_9(x,t)$.

4.2. Example 2

Let considerer the equation

$$w_t - w_{xx} = -4e^{-3t+x}$$

in the domain $E=(0,1)\times(0,\infty)$ and boundary condition on ∂E given by

$$w_x(0,t)=e^{-3t} \qquad w_x(1,t)=e^{1-3t}$$

$$w(x,0)=e^{x}.$$

The exact solution is $w(x,t)=e^{x-3t}$.

In Figure 3 the approximate numerical solution and the exact one are compared.Were taken $r=9$moments.

To apply Gram Schmidt to $\{K_{ij}(\tau,\xi)\}$ we consider the inner product

$$\langle f_i(\tau,\xi), f_j(\tau,\xi)\rangle = \int_{a_1}^{b_1}\int_{a_2}^{\infty} f_i(\tau,\xi) f_j(\tau,\xi) \mathrm{e}^{-\xi} \mathrm{d}\xi \mathrm{d}\tau.$$

The accuracy is, with this inner product $\|p_9(x,t)-w(x,t)\|^2 = 0.0250523.$

5. THE ONE-DIMENSIONAL ONE-PHASE INVERSE STEFAN PROBLEM

5.1. The Inverse One-Phase Stefan Problem as Integral Equation

The Stefan problem consists of finding w y s such that

$$\frac{\partial w}{\partial t} = \frac{\partial^2 w}{\partial x^2} \qquad \text{in } \{(x,t); 0 < x < s(t); t > 0\} \tag{21}$$

$$-\frac{\partial w}{\partial x}(0,t) = f(t) \qquad t > 0 \tag{22}$$

$$w(s(t),t) = 0 \qquad t > 0 \tag{23}$$

$$\frac{\mathrm{d}s}{\mathrm{d}t} = -\frac{\partial w}{\partial x}(s(t),t) \qquad t > 0 \tag{24}$$

$$w(x,0) = w_0(x) \qquad x \ge 0 \tag{25}$$

$$s(0) = a. \tag{26}$$

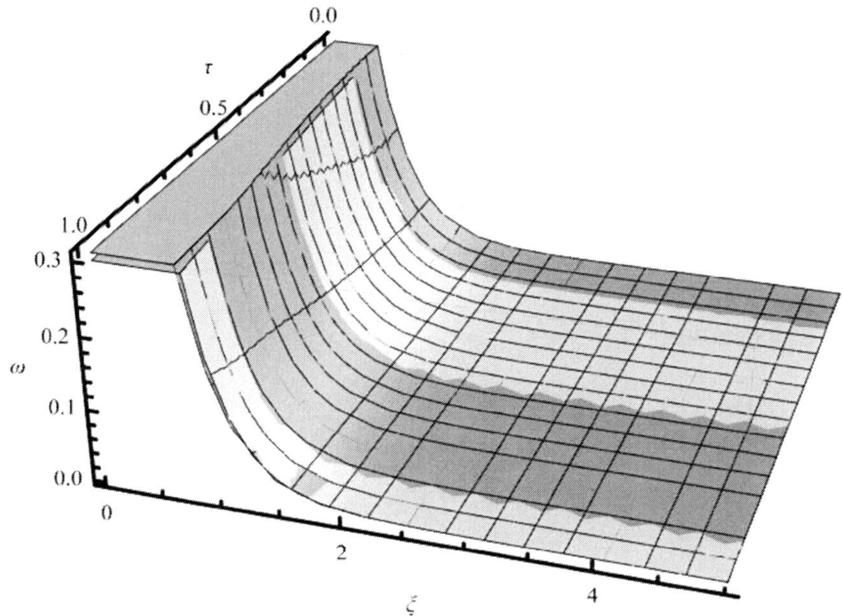

Figure 3. $w(x,t) = e^{x-3t}$ and $p_9(x,t)$.

We want to solve the inverse Stefan problem: to find $f(t)$ with $s(t)$ known such that the above con- ditions are met.

We write

$$w_\xi - (w_\tau)_\tau = 0. \tag{27}$$

We take the auxiliary function

$$u(x,t,\tau,\xi) = e^{-(t+1)\xi} \cos(x\tau).$$

Therefore

$$\frac{\partial^2 u}{\partial \tau^2} = u_{\tau\tau} = x^2 u \qquad \frac{\partial u}{\partial \xi} = u_\xi = -(t+1)u.$$

We consider the vector field with. In this manner $div(F^*(w)) = w_{\tau\tau} - w_\xi$. In consequence if w it is solution of the Equation (27), then $div(F^*(w)) = 0$. Reciprocally, if w satisfies $div(F^*(w)) = 0$, then w it is solution of the Equation (27).

We write $E = \{(\tau,\xi); 0 < \tau < s(\xi); \xi > 0\}$. We use that

$$div(uF^*) = udiv(F^*) - F^* \cdot \nabla u$$

if $\cdot$ it is the scalar product and ∇ is the gradient operator we get

$$\iint_E udiv(F^*)dA = \iint_E div(uF^*)dA - \iint_E F^* \cdot \nabla u dA.$$

By the divergence theorem

$$\iint_E div(uF^*)dA = \int_C (uF^*) \cdot nds$$

with $C = \partial E$, in consequence

$$\iint_E udiv(F^*)dA = \int_C (uF^*) \cdot nds - \iint_E F^* \cdot \nabla u dA.$$

We calculate $\iint_E F^* \cdot \nabla u dA$:

First we write

$$F^* \cdot \nabla u = (F_1, F_2) \cdot (u_\tau, u_\xi) = F_1 u_\tau + F_2 u_\xi$$

and $E_T = \{(\tau,\xi); 0<\tau<s(\xi); 0<\xi<T\}$, (Figure 4) then we take $T\to\infty$.

If $F_1^p(w) = \int F_1(w)\,d\tau = w$ then

$$\iint_{E_T} F_1 u_\tau dA = \int_0^T\int_0^{s(\xi)} F_1 u_\tau d\tau d\xi$$
$$= \int_0^T\left[F_1^p\left(w\left(s(\xi),\xi\right)\right)u_\tau\left(x,t,s(\xi),\xi\right) - F_1^p(0,\xi)u_\tau(x,t,0,\xi)\right]d\xi - \int_0^T\int_0^{s(\xi)} u_{\tau\tau}F_1^p d\tau d\xi.$$

As $w(s(\xi),\xi)=0$ by (23)

and $u_\tau(x,t,0,\xi) = x\sin(x\cdot 0)e^{-(t+1)\xi} = 0$ then

$$\iint_{E_T} F^\bullet\cdot\nabla u dA = -\int_0^T\int_0^{s(\xi)}\left(u_{\tau\tau}F_1^p(w) - F_2(w)u_\xi\right)d\tau d\xi = \int_0^T\int_0^{s(\xi)} uw\left(x^2+(t+1)\right)d\tau d\xi. \tag{28}$$

Now we developed $\int_c (uF^\bullet)\cdot n ds$. Observe the Figure 5 and:

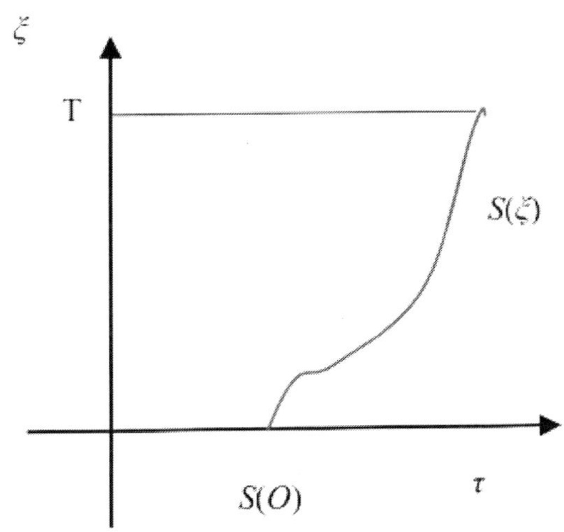

Figure 4. Domain E_T.

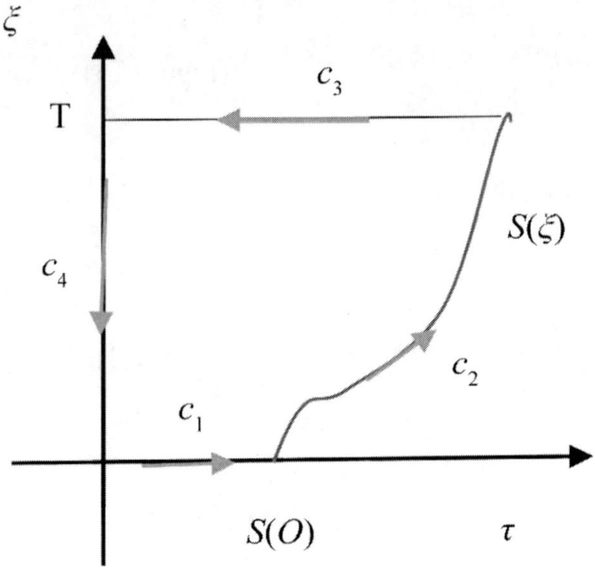

Figure 5. Domain E_T.

$$\int_{c1}\left(uF^{*}\right)\cdot n\mathrm{d}s=\int_{0}^{s(0)}u\left(x,t,\tau,0\right)w\left(\tau,0\right)\mathrm{d}\tau=\int_{0}^{s(0)}u\left(x,t,\tau,0\right)w_{0}\left(\tau\right)\mathrm{d}\tau \tag{29}$$

$$\int_{c3}\left(uF^{*}\right)\cdot n\mathrm{d}s=\int_{0}^{s(T)}u\left(x,t,\tau,T\right)w\left(\tau,T\right)\mathrm{d}x\xrightarrow[T\to\infty]{}0 \tag{30}$$

$$\int_{c4}\left(uF^{*}\right)\cdot n\mathrm{d}s=-\int_{0}^{T}u\left(x,t,0,\xi\right)w_{\tau}\left(0,\xi\right)\mathrm{d}\xi \tag{31}$$

$$\int_{c2}\left(uF^{*}\right)\cdot n\mathrm{d}s=-\int_{0}^{T}u\left(x,t,s\left(\xi\right),\xi\right)s_{\xi}\left(\xi\right)\mathrm{d}\xi. \tag{32}$$

Then

$$\int_{0}^{\infty}\int_{0}^{s(\xi)}uw\left(x^{2}+\left(t+1\right)\right)\mathrm{d}\tau\mathrm{d}\xi$$
$$=\int_{0}^{s(0)}\cos\left(x\tau\right)w\left(\tau,0\right)\mathrm{d}\tau-\int_{0}^{\infty}\cos\left(xs\left(\xi\right)\right)\mathrm{e}^{-(t+1)\xi}s_{\xi}\left(\xi\right)\mathrm{d}\xi-\int_{0}^{\infty}\mathrm{e}^{-(t+1)\xi}w_{\tau}\left(0,\xi\right)\mathrm{d}\xi. \tag{33}$$

To solve the inverse problem, where $s(\xi)$ is known and $w_\tau(0,\xi)$ is unknown we do $x^2+t+1=0$.

In this manner:

$$\int_0^\infty e^{-(t+1)\xi} w_\tau(0,\xi)\,d\xi$$
$$= \int_0^{s(0)} \cos\left(\sqrt{-t-1}\tau\right) w(\tau,0)\,d\tau - \int_0^\infty \cos\left(\sqrt{-t-1}s(\xi)\right) e^{-(t+1)\xi} s_\xi(\xi)\,d\xi = \varphi(t).$$

(34)

We assign values to t: $t=0,1,2,\cdots$

$$\int_0^\infty e^{-(t+1)\xi} w_\tau(0,\xi)\,d\xi = \varphi(t) = \mu_t. \qquad (35)$$

We can interpret (35) as a one-dimensional generalized moments problem.

We solve the problem numerically considering the finite generalized moments problem

$$\int_0^\infty e^{-(i+1)\xi} w_\tau(0,\xi)\,d\xi = \varphi(i) = \mu_i \qquad i=0,1,\cdots,n \qquad (36)$$

and we apply the truncated expansion method.

5.2. Numerical Approximation to the Solution of the Inverse Stefan Problem

To obtain a numerical approximation of the solution the procedure is analogous to that presented in Section 3. To approximate $w_\tau(0,\xi)$ is taken a base $\phi_i(\xi), i=0,1,2,\cdots,n$ of $L^2(0,\infty)$ obtained from the sequence $g_i(\xi)=e^{-(i+1)\xi}$ by Gram-Schmidt method and necessary functions are added in order to have an orthonormal basis. We then approximate the solution $w_r(0,\xi)$ with:

$$p_n(\xi)=\sum_{i=0}^{n}\lambda_i\phi_i(\xi)$$

where

$$\lambda_i=\sum_{j=0}^{i}C_{ij}\mu_j \qquad i=0,1,\cdots,n$$

and the coefficients C_{ij} verifies

$$C_{ij}=\left(\sum_{r=j}^{i-1}(-1)\frac{\langle g_i(\xi)\,|\,\phi_r(\xi)\rangle}{\|\phi_r(\xi)\|^2}C_{rj}\right)\cdot\|\phi_i(\xi)\|^{-1} \quad 1<i\le n;1\le j<i \tag{37}$$

$$C_{ii}=\|\phi_i(\xi)\|^{-1} \qquad i=0,1,\cdots,n. \tag{38}$$

The following theorem is the one-dimensional version of Theorem 1. In [15] is the demonstration when the domain is bounded.

We present here the demonstration when the domain is the interval $(0,\infty)$.

Theorem 2. Let $\{\mu_i\}_{i=0}^{n}$ be a set of real numbers and let ε and M be two positive numbers such that

$$\sum_{i=0}^{n}\left|\int_0^{\infty}g_i(\xi)x(\xi)\mathrm{d}\xi-\mu_i\right|^2\le\varepsilon^2 \tag{39}$$

$$\int_0^{\infty}\left[\xi e^{\xi}x_{\xi}^2(\xi)\right]\mathrm{d}\xi\le M^2 \tag{40}$$

then if $\lim_{\xi\to\infty}\xi^r x(\xi)=0, \forall r\in N$

$$\int_0^\infty |x(\xi)|^2\,d\xi \le \min_n\left\{\|C^T C\|\varepsilon^2+\frac{M^2}{(i+1)}; i=0,1,\cdots,n\right\} \tag{41}$$

And

$$\int_0^\infty |x(\xi)-p_n(\xi)|^2\,d\xi \le \|C^T C\|\varepsilon^2+\frac{M^2}{(n+1)} \tag{42}$$

where C is the triangular matriz with elements $C_{ij}\ (1<i\le n; 1\le j<i)$.

Proof. Since the problem is linear we can assume $\mu_i=0, i=0,1,\cdots,n$.

We applied Gram-Scmidt method on $\{g_i(\xi)\}_{i=0}^n$ in $L^2(0,\infty)$ and we get $\{\phi_i(\xi)\}_{i=0}^\infty$ then add the resulting set of necessary functions to obtain an orthonormal basis.

We write $x(\xi)$ as

$$x(\xi)=h_n(\xi)+f_n(\xi)$$

where $h_n(\xi)$ it is the orthogonal projection of $x(\xi)$ on the linear space generated by the set $\{g_i(\xi)\}_{i=0}^n$ and $f_n(\xi)=x(\xi)-h_n(\xi)$ it is the orthogonal projection of $x(\xi)$ on the orthogonal complement. Here the underlying structure is the space $L^2(0,\infty)$. We can write

$$h_n(\xi)=\sum_{i=0}^n \lambda_i\phi_i(\xi) \qquad f_n(\xi)=\sum_{i=n+1}^\infty \lambda_i\phi_i(\xi) \tag{43}$$

where $\lambda_i = \int_0^\infty \phi_i(\xi)x(\xi)\mathrm{d}\xi, i = 0,1,\cdots$ are the Fourier coefficients in the expansion of $x(\xi)$.

To estimate $h_n(\xi)$ we consider the relationship between the Fourier coefficients λ_i and the moments

$$\mu_i = \int_0^\infty g_i(\xi)x(\xi)\mathrm{d}t$$

:

$$\lambda_i = \sum_{j=0}^{i} C_{ij}\mu_j \qquad i = 0,1,\cdots,n \tag{44}$$

where C_{ij} they are given in (37) y (38).

In matrix notation

$$\lambda = \begin{pmatrix} \lambda_0 \\ \lambda_1 \\ \vdots \\ \lambda_n \end{pmatrix} \qquad \mu = \begin{pmatrix} \mu_0 \\ \mu_1 \\ \vdots \\ \mu_n \end{pmatrix}. \tag{45}$$

Then

$$\sum_{j=0}^{i} C_{ij}g_j(\xi) = \phi_i(\xi) \qquad i = 0,1,\cdots,n. \tag{46}$$

By (43) until (46) we can write

$$\int_0^\infty |h_n(\xi)|^2 \mathrm{d}\xi = \langle\lambda|\lambda\rangle = \langle C^\mathrm{T}C\mu|\mu\rangle \le \|C^\mathrm{T}C\|\|\mu\|^2 \le \|C^\mathrm{T}C\|\varepsilon^2. \tag{47}$$

To estimate f_n we see that each element of the orthonormal set $\phi_i(t)$ can be expanded in terms of the elements other orthonormal basis, in particular the base $\left\{l_i(\xi)=L_i(\xi)\mathrm{e}^{-\xi}\right\}_{i=0}^{\infty}$, where $L_i(\xi)$ it represents the Laguerre polynomial of degree i.

These polynomials satisfy

$$\xi\frac{\mathrm{d}^2}{\mathrm{d}\xi^2}\left[L_i(\xi)\right]+(1-\xi)\frac{\mathrm{d}}{\mathrm{d}\xi}\left[L_i(\xi)\right]+iL_i(\xi)=0 \tag{48}$$

or also

$$-\frac{\mathrm{d}}{\mathrm{d}\xi}\left[\xi\frac{\mathrm{d}}{\mathrm{d}\xi}\left[L_i(\xi)\right]\right]+\xi\frac{\mathrm{d}}{\mathrm{d}\xi}\left[L_i(\xi)\right]=iL_i(\xi). \tag{49}$$

Then

$$\phi_i(\xi)=\sum_{j=0}^{\infty}\gamma_{ij}l_j(\xi) \qquad i=0,1,2,\cdots,n \tag{50}$$

then using (50)

$$f_n(\xi)=\sum_{i=n+1}^{\infty}\lambda_i\phi_i(\xi)=\sum_{i=n+1}^{\infty}\sum_{j=0}^{\infty}\lambda_i\gamma_{ij}l_j(\xi). \tag{51}$$

After several calculations

$$\int_0^{\infty}\left|f_n(\xi)\right|^2\mathrm{d}\xi=\left\|f_n(\xi)\right\|^2=\sum_{i=n+1}^{\infty}\left|\sum_{j=0}^{\infty}\lambda_j\gamma_{ji}\right|^2 \tag{52}$$

and

$$\int_0^{\infty}\left|f_n(\xi)\right|^2\mathrm{d}\xi\le\frac{1}{n+1}\sum_{i=0}^{\infty}(i+1)\left|\chi(i)\right|^2 \tag{53}$$

where

$$\chi(i)=\sum_{j=0}^{\infty}\lambda_j\gamma_{ji}. \tag{54}$$

Now multiplying by $x(\xi)$ and integrating both sides of the differential Equation (49) and assuming that $\lim_{\xi\to\infty}\xi^r x(\xi)=0, \forall r\in N$ we get:

$$\int_0^{\infty}\xi x_\xi^2(\xi)e^{\xi}d\xi=\sum_{i=0}^{\infty}(i+1)\chi^2(i) \tag{55}$$

then by (53) and (55):

$$\int_0^{\infty}|f_n(\xi)|^2 d\xi\le\frac{1}{n+1}\int_0^{\infty}\xi x_\xi^2(\xi)e^{\xi}d\xi\le\frac{M^2}{n+1} \tag{56}$$

from (47) and (56):

$$\int_0^{\infty}|x(\xi)|^2 d\xi\le\left\|C^{T}C\right\|\varepsilon^2+\frac{M^2}{(n+1)}. \tag{57}$$

This inequality remains valid if we replace any integer i between 0 and n for n. Then the result (41) it demon- strated. An analogous demonstration proves inequality (42). □

6. NUMERICAL EXAMPLE

Find $w_\tau(0,\xi)$ such that

$$\frac{\partial w}{\partial\xi}=\frac{\partial^2 w}{\partial\tau^2}\quad\text{in }\{(\tau,\xi);0<\tau<s(\xi);\xi>0\}\quad\text{with } s(\xi)=\sqrt{1+\xi}$$

$$w(s(\xi),\xi)=0 \qquad \xi>0$$

$$\frac{ds}{d\xi}=-\frac{\partial w}{\partial \tau}(s(\xi),\xi) \qquad \xi>0$$

$$w(\tau,0)=w_0(\tau)=0.5e^{0.25}\sqrt{\pi}\left[erf(0.25)-erf\left(\frac{x}{2}\right)\right] \qquad \tau\geq 0.$$

The exact solution is $w_\tau(0,\xi)=-\frac{e^{0.25}}{2}\sqrt{\frac{1}{1+\xi}}.$

In Figure 6 the approximate numerical solution and the exact one are compared.Were taken $r=6$ moments.

To apply Gram Schmidt to $\{g_i(\xi)\}_{i=0}^{n}$ en $L^2(0,\infty)$ we considerer the inner product

$$\langle f_i(\xi),f_j(\xi)\rangle=\int_0^\infty f_i(\xi)f_j(\xi)e^{-\xi}d\xi.$$

The accuracy is, with this inner product $\|p_6(\xi)-w_\tau(0,\xi)\|^2=0.00306424.$

7. CONCLUSIONS

The parabolic partial differential equations

$$w_\xi-(w_\tau)_\tau=r(\tau,\xi)$$

on a region $E=(0,b_1)\times(a_2,b_2)$ can be written as an Fredholm integral equation

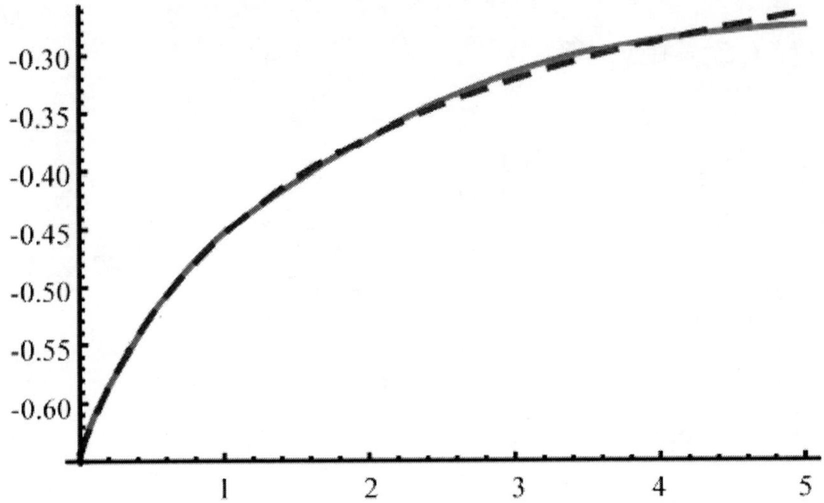

Figure 6. $w_\tau(0,\xi) = -\frac{e^{0.25}}{2}\sqrt{\frac{1}{1+\xi}}$ and $p_6(\tau,\xi)$.

$$\int_{a_2}^{b_2}\int_0^{b_1} w(\tau,\xi)u(x,t,\tau,\xi)\,d\tau d\xi = \frac{G(x,t)-R(x,t)}{\left(\left(\frac{x\pi}{b_1}\right)^2+t\right)} = \varphi(x,t) \qquad x \in N.$$

This equation is of the form:

$$\int_0^{b_1}\int_{a_2}^{b_2} w(\tau,\xi)K(x,t,\tau,\xi)\,d\tau d\xi = \varphi(x,t)$$

with $K(x,t,\tau,\xi) = e^{-t\xi}\cos\left(\frac{x\pi}{b_1}\tau\right), x \in N.$

If $w(\tau,\xi) \in L^2(E)$, then this Fredholm integral equation of first kind can be transformed into a bi-dimen- sional generalized moment problem assigning integer values greater than or equal to zero to variables x and t

$$\int_0^{b_1}\int_{a_2}^{b_2} w(\tau,\xi)K_{ij}(\tau,\xi)\mathrm{d}\tau\mathrm{d}\xi = \mu_{ij} \quad i=0,1,2,\cdots,\ j=0,1,2,\cdots. \tag{58}$$

As the functions are linearly independent then the generalized moment problem defined by (58) can be solved numerically considering the correspondent finite problem.

The inverse Stefan problem which it is to find being $w(\tau,\xi)$ unknown and such that the follow- ing conditions are met

$$\frac{\partial w}{\partial \xi} = \frac{\partial^2 w}{\partial \tau^2} \qquad \text{in } \{(\tau,\xi); 0<\tau<s(\xi); \xi>0\}$$

$$-\frac{\partial w}{\partial \tau}(0,\xi) = f(\xi) \qquad \xi>0$$

$$w(s(\xi),\xi) = 0 \qquad \xi>0$$

$$\frac{\mathrm{d}s}{\mathrm{d}\xi} = -\frac{\partial w}{\partial \tau}(s(\xi),\xi) \qquad \xi>0$$

$$w(\tau,0) = w_0(\tau) \qquad \tau\ge 0$$

$$s(0) = a$$

is equivalent to solve the integral equation

$$\int_0^{\infty} \mathrm{e}^{-(t+1)\xi} w_\tau(0,\xi)\mathrm{d}\xi$$
$$= \int_0^{s(0)} \cos\left(\sqrt{-t-1}\tau\right) w(\tau,0)\mathrm{d}\tau - \int_0^{\infty} \cos\left(\sqrt{-t-1}s(\xi)\right)\mathrm{e}^{-(t+1)\xi} s_\xi(\xi)\mathrm{d}\xi = \varphi(t)$$

which is equivalent to the generalized moments problem

$$\int_0^\infty e^{-(i+1)\xi} w_\tau(0,\xi)\,d\xi = \varphi(i) = \mu_i \qquad i = 0,1,2,\cdots$$

and can be solved numerically considering the correspondent finite problem.

REFERENCES

1. Pintarelli, M.B. and Vericat, F. (2014) Partial Differential Equations as Three-Dimensional Inverse Problem of Moments. Journal of Mathematics and System Science, 4, 657-666.

2. Cherniha, R.M. (2011) New Exact Solutions of One Nonlinear Equation in Mathematical Biology and Their Properties. Ukrainian Mathematical Journal, 53, 393-411.

3. Mittal, R.C. and Jiwari, R. (2011) A Higher Order Numerical Scheme for Some Nonlinear Differential Equations: Models in Biology. International Journal for Computational Methods in Engineering Science and Mechanics, 12, 134-140.

4. Forsythe, G.E. and Wasow, W.R. (1960) Finite Difference Methods for Partial Differential Equations. John Wiley and Sons, New York.

5. Zafarullah, A. (1971) Some Stable Implicit Difference Methods for Heat Equation with Derivative Boundary Condition. The Computer Journal, 14, 309-311.

6. Keast, P. and Mitchell, A.R. (1966) On the Instability of the Crank Nicholson Formula under Derivative Boundary Conditions. The Computer Journal, 9, 110-114.

7. Akheizer, N.I. (1965) The Classical Moment Problem. Olivier and Boyd, Edinburgh.

8. Akheizer, N.I. and Krein, M.G. (1962) Some Questions in the Theory of Moment. American Mathematical Society, Providence.

9. Ang, D.D., Gorenflo, R., Le, V.K. and Trong, D.D. (2002) Moment Theory and Some Inverse Problems in Potential Theory and Heat Conduction. Lectures Notes in Mathematics, Springer-Verlag, Berlin.

10. Shohat, J.A. and Tamarkin, J.D. (1943) The Problem of Moments. Mathematics Survey, American Mathematical Society, Providence.

11. Talenti, G. (1987) Recovering a Function from a Finite Number of Moments. Inverse Problems, 3, 501-517.

12. Pintarelli, M.B. and Vericat, F. (2011) Bi-Dimensional Inverse Moment Problems. Far East Journal of Mathematical Sciences, 54, 1-23.

13. Pintarelli, M.B. and Vericat, F. (2012) Klein-Gordon Equation as a Bi-Dimensional Moment Problem. Far East Journal of Mathematical Sciences, 70, 201-225.

14. Pintarelli, M.B. (2015) Linear Partial Differential Equations of First Order as Bi-Dimensional Inverse Moment Problem. Applied Mathematics, 6, 979-989.

15. Pintarelli, M.B. and Vericat, F. (2008) Stability Theorem and Inversion Algorithm for a Generalized Moment Problem. Far East Journal of Mathematical Sciences, 30, 253-274.

CHAPTER 4

Similarity Reduction of Nonlinear Partial Differential Equations

Amnah S. Al-Johani 1, 2

[1] Department of Applied Mathematics, College of Science, Northern Borders University, Arar, Saudi Arabia

[2] College of Home Economics, Northern Borders University, Arar, Saudi Arabia

ABSTRACT

In this work, the HB method is extended to search for similarity reduction of nonlinear partial differential equations. This method is generalized and will apply for a (2 + 1)-dimensional higher order Broer-Kaup System. Some new exact solutions of Broer-Kaup System are found.

KEYWORDS

Similarity Reduction; Exact Solutions; Nonlinear Partial Differential Equations

1. INTRODUCTION

In the past few decades, there has been the noticeable progress in the construction of the exact solutions for nonlinear partial differential equations, which has long been a major concern for both mathematicians and physicists. The effort in finding exact solutions to nonlinear differential equation, when they exist, is very important for the understanding of most nonlinear physical phenomena. For instances, the nonlinear wave phenomena observed in fluid dynamics, plasma and optical fibers are often modelled by the bell shaped sech solutions and the kink shaped tanh solutions.

We consider the following a (2 + 1)-dimensional higher order Broer-Kaup system:

$$u_t + 4\left(u_{xx} + u^3 - 3uu_x + 3uw + 3p\right)_x = 0, \tag{1}$$

$$v_t + 4\left(v_{xx} + 3vu^2 + uv_x + 3vw\right)_x = 0, \tag{2}$$

$$w_y - v_x = 0, \tag{3}$$

$$p_y - \left(uv\right)_x = 0. \tag{4}$$

which is obtained from the Kadomtsev-Petviashvili (KP) equation by the symmetry constraint [1].

The systems (1)-(4) were given by Li et al [2] solving it via a transformation and tanh-function method to obtain many new exact solutions. Jain et al. [3] reduced a system to a simple (1 + 1)-dimensional nonlinear evolution equation through a simple transformation, and by using the new generally projective Riccati equation expansion method to explore many families of soliton-like and periodic solutions for it. Recently, Li et al. [4] have obtained some new types of multisoliton solutions for the systems (1)-(4) by using some simple transformations as $v = u_y, w = u_x + \alpha(x,t), p = uu_x + \beta$ and homogenous balance method.

The homogenous balance (HB) method is a powerful tool to find solitary wave solutions of nonlinear partial differential equations. Fan et al. [5] presented an improved HB method to obtain more other kinds of exact solutions and introduced a continuation of [5] in [6]. The traditional method for finding similarity reduction of nonlinear partial differential equations is to use classical Lie approach [7,8]. However, the method involves tedious algebraic calculations and still can not be used to find all similarity solutions. Recently,Clarkson and Kruskal devloped a direct and simple method to find more similarity solutions of nonlinear PDEs.

In this work, the HB method is extended to search for similarity reduction of nonlinear partial differential equations. So, more solutions can be obtained by the improved HB method. This method is generalized and can be applied to other nonlinear partial differential equations [9-15].

Similarity Reduction of Nonlinear Partial Differential Equations

We describe the main steps of our method. For a given PDE, say in three variables, say x, y, t

$$u_t = K\left(u, u_x, u_y, u_{xx}, u_{yy}, \cdots\right) \tag{5}$$

we seek its similarity reductions in the form

$$u(x,y,t)=\frac{\partial^{\alpha}}{\partial x^{\alpha}}f(s)+u_0, \tag{6}$$

where α is a constant to determine by balancing between the highest order derivative of the linear terms of u and the nonlinear terms of u, where s, u_0 are regarded as undetermined functions.

Substituting from Equation (6) into Equation (5) and collecting all terms of f with the same derivative and power. To make the associated equation be an ordinary equations of f and s, requiring ratios of their coefficients being functions of s, we obtain a set of determining equations for s, u_0 and other undermined functions, from which s and u_0 will be obtained.

To explain this method, we will apply for a (2 + 1)-dimensional higher order Broer-Kaup system (1)-(4), we suppose their similarity solutions are of the form

$$u(x,y,t)=\frac{\partial^{\alpha_1}}{\partial x^{\alpha_1}}f(s)+u_0,\ v(x,y,t)=\frac{\partial^{\alpha_2}}{\partial x^{\alpha_2}}g(s)+v_0,$$
$$w(x,y,t)=\frac{\partial^{\alpha_3}}{\partial x^{\alpha_3}}h(s)+w_0,\ p(x,y,t)=\frac{\partial^{\alpha_4}}{\partial x^{\alpha_4}}k(s)+p_0. \tag{7}$$

Where

$$s=s(x,y,t), u_0=u_0(x,y,t), v_0=v_0(x,y,t),\ w_0=w_0(x,y,t), p_0=p(x,y,t).$$

are determined functions. Balancing the highest order of linear term with the nonlinear terms in every equations (1)-(4) to determining $\alpha_1,\alpha_2,\alpha_3,\alpha_4$, we obtain

$$\alpha_1 = 1, \alpha_2 = \alpha_3 = 2, \alpha_4 = 3 \tag{8}$$

the Equation (64) take the following form

$$u(x,y,t) = \frac{\partial}{\partial x} f(s) + u_0 = f's_x + u_0, \tag{9}$$

$$v(x,y,t) = \frac{\partial^2}{\partial x^2} g(s) + v_0 = g''s_x^2 + g's_{xx} + v_0, \tag{10}$$

$$w(x,y,t) = \frac{\partial^2}{\partial x^2} h(s) + w_0 = h''s_x^2 + h's_{xx} + w_0, \tag{11}$$

$$p(x,y,t) = \frac{\partial^3}{\partial x^3} k(s) + p_0 = k'''s_x^3 + 3k''s_x s_{xx} + k's_{xxx} + p_0. \tag{12}$$

Substituting Equations (9)-(12) into the original system (1)-(4) and collecting all terms of f, g, h, k with the same derivative and power leads to

$$\begin{aligned}
&(-12s_x s_{xxx}) f'^2 + \left(12u_{0x}s_x^2 + 36u_0 s_x s_{xx}\right)h'' + \left(36s_{xx}^2 + 48s_x s_{xx}\right)k'' + 12fh's_x s_{xxx} \\
&+\left(24u_0 u_{0x}s_x + 12w_{0x}s_x + s_{xt} + 12u_0^2 s_{xx} + 12w_0 s_{xx} - 24u_{0x}s_{xx} - 12u_0 s_{xxx} + 4s_{xxxx}\right)f' \\
&+\left(12u_{0x}s_{xx} + 12u_0 s_{xxx}\right)h' + 12k's_{xxxx} + \left(12p_{0x} + u_{0t} + 12u_0^2 u_{0x} + 12w_0 u_{0x} - 12u_{0x}^2 - 12u_0 u_{0xx} + 4u_{0xxx}\right) = 0,
\end{aligned} \tag{13}$$

$$\begin{aligned}
&(24ff''g''+12f'2g'''+4f''g'''+12h''g'''+12g''h''+4f'g''''+4g''''')s_x^5\\
&+\left(4u_0s_x^4+40s_x^3s_{xx}\right)g'''+\left(24u_0s_x^4+28s_x^3s_{xx}\right)f'g''+\left(24u_0s_x^4+12s_x^3s_{xx}\right)f''g''\\
&+\left(24f'g'f''+60f'^2g''+72g''h''+12h'g'''+12g'h'''\right)s_x^3s_{xx}\\
&+\left(s_x^2s_t+12u_0^2s_x^3+12w_0s_x^3+4u_{0x}s_x^3+24u_0s_x^2s_{xx}+60s_xs_{xx}^2+40s_x^2s_{xxx}\right)g'''\\
&+\left(24u_{0x}s_x^3+96u_0s_x^2s_{xx}+24s_xs_{xx}^2+16s_x^2s_{xxx}\right)f'g''+\left(24s_xs_{xx}^2+12s_x^2s_{xx}\right)f'^2g'\\
&+\left(24u_0s_x^2s_{xx}+4s_x^2s_{xxx}\right)f''g'+\left(36s_xs_{xx}^2+12s_x^2s_{xxx}\right)h'g''\\
&+\left(36s_xs_{xx}^2+12s_x^2s_{xxx}\right)g'h''+(24ff''+12h''')v_0s_x^3\left(24u_0v_0s_x^2+4v_{0x}s_x^2\right)f''\\
&+\left(24u_0u_{0x}s_x^2+12w_{0x}s_x^2+2s_xs_{xt}+36u_0^2s_xs_{xx}+36w_0s_xs_{xx}+12u_{0x}s_xs_{xx}+12u_0s_{xx}^2+16u_0s_xs_{xxx}\right.\\
&\left.+40s_{xx}s_{xxx}+20s_xs_{xxxx}\right)g''+\left(12v_{0x}s_x^2+36v_0s_xs_{xx}\right)h''+\left(12v_{0x}s_x^2+24v_0s_xs_{xx}\right)f'^2\\
&+\left(24u_0s_xs_{xx}+24u_0s_{xx}^2+24u_0s_xs_{xxx}+4s_{xx}s_{xxx}+4s_xs_{xxxx}\right)f'g'+24s_{xx}s_{xxxx}g'h'\\
&+\left(24v_0u_{0x}s_x+24u_0v_0s_{xx}+4v_{0x}s_{xx}\right)f'\\
&+\left(24u_0u_{0x}s_{xx}+12w_{0x}s_{xx}+s_{xxt}+12u_0^2s_{xxx}+12w_0s_{xxx}+4u_{0x}s_{xxx}+4u_0s_{xxxx}+4s_{xxxxx}\right)g'\\
&+\left(12v_{0x}s_{xx}+12v_0s_{xxx}\right)h'\\
&+\left(24u_0v_0u_{0x}+v_{0t}+12u_0^2v_{0x}+12w_0v_{0x}+4u_{0x}v_{0x}+12v_0w_{0x}+4u_0v_{0xx}+4v_{0xxx}\right)=0,
\end{aligned}\tag{14}$$

$$-g'''s_x^3+h'''s_x^2s_y+\left(s_{xx}s_y+2s_xs_{xy}\right)h''-3g's_xs_{xx}+s_{xxy}h'-s_{xxx}g'+\left(w_{0y}-v_{0x}\right)=0,\tag{15}$$

$$\begin{aligned}
&-(f''g''+f'g''')s_x^4-\left(u_{0x}s_x^2+3u_0s_xs_{xx}\right)g''-\left(s_{xx}^2+s_xs_{xxx}\right)f'g'\\
&+\left(s_ys_{xxx}+3s_{xx}s_{xy}+3s_xs_{xxy}\right)k''-v_0s_x^2f''+\left(3s_xs_ys_{xx}+3s_x^2s_{xy}+s_x^3s_y\right)k'''\\
&-(g'f''+4f'g'')s_x^2s_{xx}-u_0s_x^3g'''-\left(v_{0x}s_x+v_0s_{xx}\right)f'-\left(u_{0x}s_{xx}+u_0s_{xxx}\right)g'\\
&+s_{xxxy}k'+\left(p_{0y}-v_0u_{0x}-u_0v_{0x}\right)=0.
\end{aligned}\tag{16}$$

To make Equations (13)-(16) be an ordinary differential equations of f,g,h and k only for s, the ratios of the coefficients of different derivative and power of f,g,h,k must be functions of s. That is to say, the following constrained conditions are satisfied

$$s_x^2 s_{xx} = s_x^4 \Gamma_1(s) \tag{17}$$

$$-12u_0 s_x^3 + 24 s_x^2 s_{xx} = s_x^4 \Gamma_2(s) \tag{18}$$

$$24u_0 s_x^3 - 60 s_x^2 s_{xx} = s_x^4 \Gamma_3(s) \tag{19}$$

$$s_x s_t + 12u_0 s_x^2 + 12w_0 s_x^2 + 12 s_{xx}^2 - 24u_{0x} s_x^2 - 36u_0 s_x s_{xx} + 16 s_x s_{xxx} = s_x^4 \Gamma_4(s) \tag{20}$$

$$12u_{0x} s_x^2 + 36u_0 s_x s_{xx} = s_x^4 \Gamma_5(s) \tag{21}$$

$$36 s_{xx}^2 + 48 s_x s_{xx} = s_x^4 \Gamma_6(s) \tag{22}$$

$$12u_{0x} s_x^2 + 24u_0 s_x s_{xx} - 12 s_{xx}^2 - 12 s_x s_{xxx} = s_x^4 \Gamma_7(s) \tag{23}$$

$$s_x s_{xxx} = s_x^4 \Gamma_8(s) \tag{24}$$

$$12(w_0 s_x)_x + 12(u_0^2 s_x)_x + s_{xt} - 24u_{0x} s_{xx} - 12u_0 s_{xxx} + 4 s_{xxxx} = s_x^4 \Gamma_9(s) \tag{25}$$

$$12u_{0x} s_{xx} + 12u_0 s_{xxx} = s_x^4 \Gamma_{10}(s) \tag{26}$$

$$s_{xxxx} = s_x^4 \Gamma_{11}(s) \tag{27}$$

$$12p_{0x} + u_{0t} + 12u_0^2 u_{0x} + 12w_0 u_{0x} - 12u_{0x}^2 - 12u_0 u_{0xx} + 4u_{0xxx} = s_x^4 \Gamma_{12}(s) \tag{28}$$

$$s_{xx}s_x^3 = s_x^4\Gamma_{13}(s) \tag{29}$$

$$4u_0 s_x^4 + 40 s_x^3 s_{xx} = s_x^5\Gamma_{14}(s) \tag{30}$$

$$24u_0 s_x^4 + 28 s_x^3 s_{xx} = s_x^5\Gamma_{15}(s) \tag{31}$$

$$24u_0 s_x^4 + 12 s_x^3 s_{xx} = s_x^5\Gamma_{16}(s) \tag{32}$$

$$s_x^2 s_t + 12u_0^2 s_x^3 + 12w_0 s_x^3 + 4u_{0x}s_x^3 + 24u_0 s_x^2 s_{xx} + 60 s_x s_{xx}^2 \tag{33}$$

$$s_x^2 s_{xx} = s_x^4\Gamma_1(s) \tag{34}$$

$$-12u_0 s_x^3 + 24 s_x^2 s_{xx} = s_x^4\Gamma_2(s) \tag{35}$$

$$24u_0 s_x^3 - 60 s_x^2 s_{xx} = s_x^4\Gamma_3(s) \tag{36}$$

$$s_x s_t + 12u_0 s_x^2 + 12w_0 s_x^2 + 12 s_{xx}^2 - 24u_{0x}s_x^2 - 36u_0 s_x s_{xx} + 16 s_x s_{xxx} = s_x^4\Gamma_4(s) \tag{37}$$

$$12u_{0x}s_x^2 + 36u_0 s_x s_{xx} = s_x^4\Gamma_5(s) \tag{38}$$

$$36 s_{xx}^2 + 48 s_x s_{xx} = s_x^4\Gamma_6(s) \tag{39}$$

$$12u_{0x}s_x^2 + 24u_0 s_x s_{xx} - 12 s_{xx}^2 - 12 s_x s_{xxx} = s_x^4\Gamma_7(s) \tag{40}$$

$$s_x s_{xxx} = s_x^4\Gamma_8(s) \tag{41}$$

$$12\left(w_0 s_x\right)_x + 12\left(u_0^2 s_x\right)_x + s_{xt} - 24u_{0x}s_{xx} - 12u_0 s_{xxx} + 4s_{xxxx} = s_x^4\Gamma_9(s) \tag{42}$$

$$12u_{0x}s_{xx} + 12u_0 s_{xxx} = s_x^4\Gamma_{10}(s) \tag{43}$$

$$s_{xxxx} = s_x^4\Gamma_{11}(s) \tag{44}$$

$$12p_{0x} + u_{0t} + 12u_0^2 u_{0x} + 12w_0 u_{0x} - 12u_{0x}^2 - 12u_0 u_{0xx} + 4u_{0xxx} = s_x^4\Gamma_{12}(s) \tag{45}$$

$$s_{xx}s_x^3 = s_x^4\Gamma_{13}(s) \tag{46}$$

$$4u_0 s_x^4 + 40s_x^3 s_{xx} = s_x^5\Gamma_{14}(s) \tag{47}$$

$$24u_0 s_x^4 + 28s_x^3 s_{xx} = s_x^5\Gamma_{15}(s) \tag{48}$$

$$24u_0 s_x^4 + 12s_x^3 s_{xx} = s_x^5\Gamma_{16}(s) \tag{49}$$

$$s_x^2 s_t + 12u_0^2 s_x^3 + 12w_0 s_x^3 + 4u_{0x}s_x^3 + 24u_0 s_x^2 s_{xx} + 60s_x s_{xx}^2 \tag{50}$$

$$+40s_x^2 s_{xxx} s_x^5 = \Gamma_{17}(s) \tag{51}$$

$$24u_{0x}s_x^3 + 96u_0 s_x^2 s_{xx} + 24s_x s_{xx}^2 + 16s_x^2 s_{xxx} = s_x^5\Gamma_{18}(s) \tag{52}$$

$$24s_x s_{xx}^2 + 12s_x^2 s_{xx} = s_x^5\Gamma_{19}(s) \tag{53}$$

$$24u_0 s_x^2 s_{xx} + 4s_x^2 s_{xxx} = s_x^5\Gamma_{20}(s) \tag{54}$$

$$36s_x s_{xx}^2 + 12s_x^2 s_{xxx} = s_x^5 \Gamma_{21}(s) \tag{55}$$

$$36s_x s_{xx}^2 + 12s_x^2 s_{xxx} = s_x^5 \Gamma_{22}(s) \tag{56}$$

$$v_0 s_x^3 = s_x^5 \Gamma_{23}(s) \tag{57}$$

$$24u_0 v_0 s_x^2 + 4v_{0x} s_x^2 = s_x^5 \Gamma_{24}(s) \tag{58}$$

$$24u_0 u_{0x} s_x^2 + 12w_{0x} s_x^2 + 2s_x s_{xt} + 36u_0^2 s_x s_{xx} + 36w_0 s_x s_{xx} + 12u_{0x} s_x s_{xx} \tag{59}$$

$$+12u_0 s_{xx}^2 + 16u_0 s_x s_{xxx} + 40s_{xx} s_{xxx} + 20s_x s_{xxxx} = s_x^5 \Gamma_{25}(s) \tag{60}$$

$$12v_{0x} s_x^2 + 36v_0 s_x s_{xx} = s_x^5 \Gamma_{26}(s) \tag{61}$$

$$12v_{0x} s_x^2 + 24v_0 s_x s_{xx} = s_x^5 \Gamma_{27}(s) \tag{62}$$

$$24u_0 s_x s_{xx} + 24u_0 s_{xx}^2 + 24u_0 s_x s_{xxx} + 4s_{xx} s_{xxx} + 4s_x s_{xxxx} = s_x^5 \Gamma_{28}(s) \tag{63}$$

$$24s_{xx} s_{xxxx} = s_x^5 \Gamma_{29}(s) \tag{64}$$

$$24v_0 u_{0x} s_x + 24u_0 v_0 s_{xx} + 4v_{0x} s_{xx} = s_x^5 \Gamma_{30}(s) \tag{65}$$

$$12\left(u_0^2 s_{xx}\right)_x + s_{xxt} + 12\left(w_0 s_{xx}\right)_x + \left(4u_0 s_{xxx}\right)_x + 4s_{xxxxx} s_x^5 = \Gamma_{31}(s) \tag{66}$$

$$12v_{0x}s_{xx}+12v_0s_{xxx}=s_x^5\Gamma_{32}(s) \tag{67}$$

$$24u_0v_0u_{0x}+v_{0t}+12u_0^2v_{0x}+12(w_0v_0)_x+4(u_0v_{0x})_x+4v_{0xxx}=s_x^5\Gamma_{33}(s) \tag{68}$$

$$-s_x^2s_y=s_x^3\Gamma_{34}(s) \tag{69}$$

$$-(s_{xx}s_y+2s_xs_{xy})=s_x^3\Gamma_{35}(s) \tag{70}$$

$$3s_xs_{xx}=s_x^3\Gamma_{36}(s) \tag{71}$$

$$-s_{xxy}=s_x^3\Gamma_{37}(s) \tag{72}$$

$$s_{xxx}=s_x^3\Gamma_{38}(s) \tag{73}$$

$$v_{0x}-w_{0y}=s_x^3\Gamma_{39}(s) \tag{74}$$

$$u_{0x}s_x^2+3u_0s_xs_{xx}=s_x^4\Gamma_{40}(s) \tag{75}$$

$$s_{xx}^2+s_xs_{xxx}=s_x^4\Gamma_{41}(s) \tag{76}$$

$$-(s_ys_{xxx}+3s_{xx}s_{xy}+3s_xs_{xxy})=s_x^4\Gamma_{42}(s) \tag{77}$$

$$v_0s_x^2=s_x^4\Gamma_{43}(s) \tag{78}$$

$$-\left(3s_x s_y s_{xx} + 3s_x^2 s_{xy} + s_x^3 s_y\right) = s_x^4 \Gamma_{44}(s) \tag{79}$$

$$s_x^2 s_{xx} = s_x^4 \Gamma_{45}(s) \tag{80}$$

$$u_0 s_x^3 = s_x^4 \Gamma_{46}(s) \tag{81}$$

$$v_{0x} s_x + v_0 s_{xx} = s_x^4 \Gamma_{47}(s) \tag{82}$$

$$u_{0x} s_{xx} + u_0 s_{xxx} = s_x^4 \Gamma_{48}(s) \tag{83}$$

$$-s_{xxxy} = s_x^4 \Gamma_{49}(s) \tag{84}$$

$$-\left(p_{0y} - v_0 u_{0x} - u_0 v_{0x}\right) = s_x^4 \Gamma_{50}(s) \tag{85}$$

where $\Gamma_i\,(i = 1,\cdots,50)$ are some arbitrary functions of s to be determined and take the following form

$$\Gamma_1(s) = \frac{12f'^2 f'' - 12f''^2 + 12f'h'' - 12ff''' + 12fh''' + 4f'''' + 12k''''}{12\left(f'^3 + h'f'' + 4fh''\right)}$$

$$\Gamma_2(s) = \frac{12f'^2 f'' - 12f''^2 + 12f'h'' - 12ff''' + 12fh''' + 4f'''' + 12k''''}{f'''}$$

$$\Gamma_3(s) = \frac{12f'^2 f'' - 12f''^2 + 12f'h'' - 12ff''' + 12fh''' + 4f'''' + 12k''''}{ff''}$$

$$\Gamma_4(s) = \frac{12f'^2 f'' - 12f''^2 + 12f'h'' - 12ff''' + 12fh''' + 4f'''' + 12k''''}{f''}$$

$$\Gamma_5(s)=\frac{12f'^2f''-12f''^2+12f''h''-12f'f'''+12f'h'''+4f''''+12k''''}{h''}$$

$$\Gamma_6(s)=\frac{12f'^2f''-12f''^2+12f''h''-12f'f'''+12f'h'''+4f''''+12k''''}{k''}$$

$$\Gamma_7(s)=\frac{12f'^2f''-12f''^2+12f''h''-12f'f'''+12f'h'''+4f''''+12k''''}{f'^2}$$

$$\Gamma_8(s)=\frac{12f'^2f''-12f''^2+12f''h''-12f'f'''+12f'h'''+4f''''+12k''''}{12f'h'}$$

$$\Gamma_9(s)=\frac{12f'^2f''-12f''^2+12f''h''-12f'f'''+12f'h'''+4f''''+12k''''}{f'}$$

$$\Gamma_{10}(s)=\frac{12f'^2f''-12f''^2+12f''h''-12f'f'''+12f'h'''+4f''''+12k''''}{h'}$$

$$\Gamma_{11}(s)=\frac{12f'^2f''-12f''^2+12f''h''-12f'f'''+12f'h'''+4f''''+12k''''}{k'}$$

$$\Gamma_{12}(s)=12f'^2f''-12f''^2+12f''h''-12f'f'''+12f'h'''+4f''''+12k''''$$

$$\Gamma_{13}(s)=\frac{24f'f''g''+12f'2g'''+4f''g'''+12h''g'''+12g''h''+4f'g''''+4g''''}{24f'g'f''+60f'^2g''+72g''h''+12h'g'''+12g'h'''}$$

$$\Gamma_{14}(s)=\frac{24f'f''g''+12f'2g'''+4f''g'''+12h''g'''+12g''h''+4f'g''''+4g''''}{g''''}$$

$$\Gamma_{15}(s)=\frac{24f'f''g''+12f'2g'''+4f''g'''+12h''g'''+12g''h''+4f'g''''+4g''''}{f'g'''}$$

$$\Gamma_{16}(s)=\frac{24f'f''g''+12f'2g'''+4f''g'''+12h''g'''+12g''h''+4f'g''''+4g''''}{f''g''}$$

$$\Gamma_{17}(s)=\frac{24ff''g''+12f'2g'''+4f''g'''+12h''g'''+12g''h''+4f'g''''+4g''''}{g'''}$$

$$\Gamma_{18}(s)=\frac{24ff''g''+12f'2g'''+4f''g'''+12h''g'''+12g''h''+4f'g''''+4g''''}{f'g''}$$

$$\Gamma_{19}(s)=\frac{24ff''g''+12f'2g'''+4f''g'''+12h''g'''+12g''h''+4f'g''''+4g''''}{f'^{2}g'}$$

$$\Gamma_{20}(s)=\frac{24ff''g''+12f'2g'''+4f''g'''+12h''g'''+12g''h''+4f'g''''+4g''''}{f''g'}$$

$$\Gamma_{21}(s)=\frac{24ff''g''+12f'2g'''+4f''g'''+12h''g'''+12g''h''+4f'g''''+4g''''}{g''h'}$$

$$\Gamma_{22}(s)=\frac{24ff''g''+12f'2g'''+4f''g'''+12h''g'''+12g''h''+4f'g''''+4g''''}{g'h''}$$

$$\Gamma_{23}(s)=\frac{24ff''g''+12f'2g'''+4f''g'''+12h''g'''+12g''h''+4f'g''''+4g''''}{\left(24ff''+12h'''\right)}$$

$$\Gamma_{24}(s)=\frac{24ff''g''+12f'2g'''+4f''g'''+12h''g'''+12g''h''+4f'g''''+4g''''}{f''}$$

$$\Gamma_{25}(s)=\frac{24ff''g''+12f'2g'''+4f''g'''+12h''g'''+12g''h''+4f'g''''+4g''''}{g''}$$

$$\Gamma_{26}(s)=\frac{24ff''g''+12f'2g'''+4f''g'''+12h''g'''+12g''h''+4f'g''''+4g''''}{h''}$$

$$\Gamma_{27}(s)=\frac{24ff''g''+12f'2g'''+4f''g'''+12h''g'''+12g''h''+4f'g''''+4g''''}{f'^{2}}$$

$$\Gamma_{28}(s)=\frac{24ff''g''+12f'2g'''+4f''g'''+12h''g'''+12g''h''+4f'g''''+4g''''}{f'g'}$$

$$\Gamma_{29}(s)=\frac{24ff'f''g''+12f'2g'''+4f''g'''+12h''g'''+12g''h''+4f'g''''+4g''''}{g'h'}$$

$$\Gamma_{30}(s)=\frac{24ff'f''g''+12f'2g'''+4f''g'''+12h''g'''+12g''h''+4f'g''''+4g''''}{f'}$$

$$\Gamma_{31}(s)=\frac{24ff'f''g''+12f'2g'''+4f''g'''+12h''g'''+12g''h''+4f'g''''+4g''''}{g'}$$

$$\Gamma_{32}(s)=\frac{24ff'f''g''+12f'2g'''+4f''g'''+12h''g'''+12g''h''+4f'g''''+4g''''}{h'}$$

$$\Gamma_{33}(s)=24ff'f''g''+12f'^{2}g'''+4f''g'''+12h''g'''+12g''h''+4f'g''''+4g''''$$

$$\Gamma_{39}(s)=g'''\ \Gamma_{38}(s)=\frac{g'''}{g'}\ \Gamma_{37}(s)=\frac{g'''}{h'}\ \Gamma_{36}(s)=\frac{g'''}{g''}$$
$$\Gamma_{35}(s)=\frac{g'''}{h''}\ \Gamma_{34}(s)=\frac{g'''}{h'''}$$

$$\Gamma_{43}(s)=\frac{f''g''+f'g'''}{f''}\ \Gamma_{42}(s)=\frac{f''g''+f'g'''}{k''}$$
$$\Gamma_{41}(s)=\frac{f''g''+f'g'''}{f'g'}\ \Gamma_{40}(s)=\frac{f''g''+f'g'''}{g''}$$

$$\Gamma_{47}(s)=\frac{f''g''+f'g'''}{f'}\ \Gamma_{46}(s)=\frac{f''g''+f'g'''}{g'''}$$
$$\Gamma_{45}(s)=\frac{f''g''+f'g'''}{g'f''+4f'g''}\ \Gamma_{44}(s)=\frac{f''g''+f'g'''}{k'''}$$

$$\Gamma_{50}(s)=f''g''+f'g'''\ \Gamma_{49}(s)=\frac{f''g''+f'g'''}{k'}\ \Gamma_{48}(s)=\frac{f''g''+f'g'''}{g'}$$

(86)

There are freedoms in the determination of u_0, v_0, w_0, p_0, s which can exploit the following rules, without loss of generality:

If u_0 has the form $u_0 = u'(x,y,t) + \frac{\partial}{\partial x}\Omega$ then we can assume that $\Omega = 0$

$\left(\text{make the transformation } f(s) \to f(s) - \Omega\right)$,

If v_0 has the form $v_0 = v'(x,y,t) + \frac{\partial^2}{\partial x^2}\Omega$ then we can assume that $\Omega = 0$

$\left(\text{make the transformation } f(s) \to f(s) - \Omega\right)$,

If w_0 has the form $w_0 = w'(x,y,t) + \frac{\partial}{\partial x}\Omega$ then we can assume that $\Omega = 0$

$\left(\text{make the transformation } f(s) \to f(s) - \Omega\right)$,

If p_0 has the form $p_0 = p'(x,y,t) + \frac{\partial^3}{\partial x^3}\Omega$ then we can assume that $\Omega = 0$

$\left(\text{make the transformation } f(s) \to f(s) - \Omega\right)$.

If $s(x,y,t)$ is defined by an equation of the form $\Omega(s) = s_0(x,y,t)$, we can also assume that $\Omega = s$ (make the transformation $s \to \Omega^{-1}(s)$) •

From Equation (17), we get

$$\frac{s_{xx}}{s_x} = s_x \Gamma_1(s) \tag{87}$$

integrating Equation (87) with respect to x, we get

$$Lns_x + Ln\Gamma(s) = Ln\theta(y,t), \tag{88}$$

$$s_x \Gamma(s) = \theta(y,t). \tag{89}$$

integrating Equation (89) with respect to x, we obtain

$$\gamma(s) = \theta(y,t)x + \sigma(y,t). \tag{90}$$

By using the rule (e) into Equation (90), we obtain a function $s(x,y,t)$ in the form

$$s(x,y,t) = \theta(y,t)x + \sigma(y,t) \tag{91}$$

substituting from Equation (90) into Equations (17), (20), (24), (25), (26), (53), (54), (55), (56), (63), (64), (71), (72), (73), (76), (77), (80), (83)and (84) we obtain

$$\begin{aligned}
&\Gamma_1(s) = \Gamma_6(s) = \Gamma_8(s) = \Gamma_{10}(s) = \Gamma_{11}(s) = \Gamma_{19}(s) = \Gamma_{20}(s) = \Gamma_{21}(s) = \Gamma_{22}(s) \\
&\Gamma_{28}(s) = \Gamma_{29}(s) = \Gamma_{31}(s) = \Gamma_{32}(s) = \Gamma_{36}(s) = \Gamma_{37}(s) = \Gamma_{38}(s) = \Gamma_{41}(s) = \Gamma_{42}(s) \\
&\Gamma_{45}(s) = \Gamma_{48}(s) = \Gamma_{49}(s) = 0,
\end{aligned} \tag{92}$$

where $s_{xx} = 0$ which clear in equation (91) .

By using Equation (91) into Equation (18), we get

$$u_0 = -\frac{1}{12}\theta\Gamma_2(s) = -\frac{1}{12}\frac{\partial}{\partial x}\left[\gamma_2(s)\right] \tag{93}$$

where $\Gamma_2(s) = \frac{d}{ds}\gamma_2(s),\ \theta = \frac{\partial s}{\partial x}$.

By apply the rule (a) on Equation (93), we obtain

$$u_0 = 0.\ \gamma_2(s) = 0 \Rightarrow \Gamma_2(s) = 0, \tag{94}$$

substituting from Equations (91), (94) into Equation (20), we obtain

$$\Gamma_3(s) = 0. \tag{95}$$

using Equations (91)and (94) into Equation (19), we get

$$w_0 = \frac{1}{12}\theta^2\Gamma_4(s) - \frac{\theta_t}{12\theta}x - \frac{\sigma_t}{12\theta} = \frac{1}{12}\frac{\partial^2}{\partial x^2}\left[\gamma_4(s)\right] - \frac{\theta_t}{12\theta}x - \frac{\sigma_t}{12\theta}, \tag{96}$$

where $\Gamma_4(s) = \frac{d^2}{ds^2}\gamma_2(s),\ \theta^2 = \frac{\partial^2 s}{\partial x^2}$.

By using the rule (c) into the above equation (96) to become in the form

$$\gamma_4(s) = 0 \Rightarrow \Gamma_4(s) = 0,$$

then Equation (96) take the form

$$w_0 = -\frac{1}{12}\left[\theta_t x + \sigma_t\right]. \tag{97}$$

substituting from Eqs.(91), (94), (97) into Equations (20), (23), (25), (29), (30), (31), (32), (33)and (34), we obtain

$$\Gamma_5(s)=\Gamma_7(s)=\Gamma_9(s)=\Gamma_{13}(s)=\Gamma_{14}(s)=\Gamma_{15}(s)=\Gamma_{16}(s)=\Gamma_{17}(s)=\Gamma_{18}(s)=0 \tag{98}$$

from Equation (91) into Equation (57), we get

$$v_0=\theta^2\Gamma_{23}(s)=\frac{\partial^2}{\partial x^2}\left[\gamma_{23}(s)\right] \tag{99}$$

where $\Gamma_{23}(s)=\frac{\mathrm{d}^2}{\mathrm{d}s^2}\gamma_{23}(s),\ \theta^2=\frac{\partial^2 s}{\partial x^2}.$

By using the rule (b) into the above equation (99)

$$v_0=0.\ \gamma_{23}(s)=0\Rightarrow\Gamma_{23}(s)=0, \tag{100}$$

substituting from Equations.(91), (94) and (97), into Equations.(61), (62), (65), (68), (75), (78), (81)and (82) then,we obtain

$$\Gamma_{26}(s)=\Gamma_{27}(s)=\Gamma_{30}(s)=\Gamma_{33}(s)=\Gamma_{40}(s)=\Gamma_{43}(s)=\Gamma_{46}(s)=\Gamma_{47}(s)=0. \tag{101}$$

Substituting from Equations (91), (94), (97) and (101) into Equations (60), (69), (70), we obtain

$$\Gamma_{25}(s)=\frac{A}{\theta^2}=C, \tag{102}$$

$$\Gamma_{34}(s)=-(Ds+F), \tag{103}$$

$$\Gamma_{35}(s)=-2D. \tag{104}$$

where $\theta=\theta(y,t),\sigma=\sigma(y,t).$

and

$$A = \frac{\theta_t}{\theta^2}, C = \frac{A}{\theta^2}, \tag{105}$$

$$D = \frac{\theta_y}{\theta^2}, F = \frac{\sigma_y}{\theta} - D\sigma, \tag{106}$$

$$B = \frac{\sigma_t}{\theta} - A\sigma. \tag{107}$$

Using this notation, the equation (97) take the following form

$$w_0 = -\frac{1}{12}\left(As + B\right), \qquad ((108))$$

substituting Equations (97) and (100) into Equation (74), we obtain

$$\Gamma_{39}\left(s\right) = \frac{1}{12}C\left(Ds + F\right), \tag{109}$$

using any equation which we need into Equation (28), we obtain

$$12p_{0x} = \theta^4\Gamma_{12}\left(s\right) = \frac{\partial^4}{\partial x^4}\gamma_{12}\left(s\right) \tag{110}$$

where $\Gamma_{12}\left(s\right) = \frac{d^4}{ds^4}\gamma_{12}\left(s\right), \; \theta^4 = \frac{\partial^4 s}{\partial x^4}.$

By using the rule (d) after differential with respect to x into Equations (110), we obtain

$$p_0 = 0. \; \gamma_{12}\left(s\right) = 0 \Rightarrow \Gamma_{12}\left(s\right) = 0, \tag{111}$$

Substituting into Equations (9)-(12), we obtain the similarity solutions of the Broer-Kaup system Equations (1)-(4) in the form

$$p(x,y,t)=\theta^3 H(s).\ w(x,y,t)=\theta^2 R(s)-\frac{1}{12}(As+B),$$
$$v(x,y,t)=\theta^2 Q(s),\ u(x,y,t)=\theta P(s), \tag{112}$$

where

$$P(s)=f',Q(s)=g'',R(s)=h'' \text{ and } H(s)=k'''. \tag{113}$$

with $s(x,y,t)=\theta(y,t)x+\sigma(y,t)$.

Substituting from Equation (112) to obtain an ordinary differential equations from the origin system (1)-(4), we get

$$P'''-3PP''+3PR'-3P'^2+3RP'+3P^2P'+3H'=0, \tag{114}$$

$$4Q'''+24QPP'+12P^2Q'+12RQ'+4P'Q'+12QR'+4PQ''-CQ=0, \tag{115}$$

$$-12Q'-ADs+12R'Ds+12FR'+24DR-FC=0, \tag{116}$$

$$(PQ)'-3DH-FH'-DsH'=0. \tag{117}$$

where $'' \, ' \, ''=\frac{d}{ds}$.

The general solution for the variable $\theta(y,t),\sigma(y,t)$ which satisfy Equations (105)-(107) are

$$\sigma(y,t)=\frac{1}{A}\left[\frac{(BD-AF)y+c_2}{At+Dy+c_1}-B\right]\theta(y,t)=-\frac{1}{At+Dy+c_1}, \tag{118}$$

where c_1, c_2 are arbitrary constants.

There some subcases for the constants A, B, D, F

$D=F=0, A\neq 0,$ the solutions of Equations (105)-(107) are

$$\sigma(y,t)=\sigma(t)=\frac{1}{A}\left[\frac{c_5}{At+c_4}-B\right]\theta(y,t)=\theta(t)=-\frac{1}{At+c_4}, \tag{119}$$

where c_4, c_5 are arbitrary constants. In this case the equations (114)-(117) take the form

$$P'''-3PP''+3PR'-3P'^2+3RP'+3P^2P'+3H'=0, \tag{120}$$

$$4Q'''+24QPP'+12P^2Q'+12(RQ)'+4P'Q'+4PQ''-CQ=0, \tag{121}$$

$$Q'=0, \tag{122}$$

$$(PQ)'=0. \tag{123}$$

the solutions for equations (120)-(123) are

$$Q(s)=c_7,\ P(s)=\frac{c_6}{c_7}, \tag{124}$$

$$H(s)=-\frac{c_6 c}{c_7}s^2-\frac{c_6 c_8}{c_7}s+c_9.\ R(s)=\frac{1}{12}cs+c_8,$$

To obtain the solutions for the original system (1)-(4), we substituting from the equations (119), (124) into Equations (112), we get

$$u(x,y,t)=-\frac{c_6}{c_7\left(At+c_4\right)},\ v(x,y,t)=-\frac{c_7}{\left(At+c_4\right)^2},$$

$$w(x,y,t)=-\frac{cs+12c_8}{12\left(At+c_4\right)^2}-\frac{1}{12}\left(As+B\right),\ p(x,y,t)=\frac{\left(c_6cs^2+c_6c_8s-c_7c_9\right)}{c_7\left(At+c_4\right)^3}. \tag{125}$$

REFERENCES

1. H. A. Zedan, "Exact Solutions for the Generalized KdV Equation by Using Backlund Transformations," Journal of the Franklin Institute, Vol. 348, No. 8, 2011, pp. 1751-1768.

2. D.-S. Li, et al., "Solving the (2 + 1)-Dimensional Higher Order Broer-Kaup System via a Transformation and Tanh-Function mEthod," Chaos, Solitons & Fractals, Vol. 20, No. 5, 2004, pp. 1021-1025.

3. J. Mei, et al., "New Soliton-Like and Periodic Solution of (2 + 1)-Dimensional Higher Order Broer-Kaup System," Chaos, Solitons & Fractals, Vol. 22, No. 3, 2004, pp. 669-674.

4. D.-S. Li, et al., "Some New Types of Multisoliton Solutions for the (2 + 1)-Dimensional Higher-Order Broer-Kaup System," Applied Mathematics and Computation, Vol. 152, No. 3, 2004, pp. 847-853.

5. E. G. Fan and H. Q. Zhang, "A Note on the Homogeneous Balance Method," Physics Letters A, Vol. 246, No. 5, 1998, pp. 403-406.

6. E. G. Fan, "Two New Applications of the Homogeneous Balance Method," Physics Letters A, Vol. 265, No. 5-6, 2000, pp. 353-357.

7. G. Bluman, "Symmetries and Differential Equations," Springer-Verlag, New York, 1989.

8. P. J. Olver, "Applications of Lie Group to Differential Equation," Springer-Verlag, New York, 1986.

9. P. A. Clarkson, "New Similarity Solutions for the Modified Boussinesq Equation," Journal of Physics A: Mathematical and General, Vol. 22, No. 13, 1989, pp. 2355-2365.

10. Y. Yu, Q. Wang and H. Q. Zhang, "The Extended Jacobi Elliptic Function Method to Solve a Generalized Hirota-Satsuma Coupled KdV Equations," Chaos, Solitons & Fractals, Vol. 26, No. 5, 2005, pp. 1415-1421.

11. J. L. Zhang, M. L. Wang, Y. M. Wang, Z. D. Fang, "The Improved F-Expansion Method and Its Applications," Physics Letters A, Vol. 350, No. 1-2, 2006, pp. 103-109.

12. E. M. E. Zayed and H. Zedan, "On the Solitary Wave Solutions for Nonlinear Hirota-Satsuma Coupled KdV of Equations," Chaos, Solitons & Fractals, Vol. 22, No. 2, 2004, pp. 285-303.

13. M. Wadati, "Introduction to Solitons," Pramana: Journal of Physics, Vol. 57, No. 5-6, 2001, pp. 841-847.

14. Z. Chen, D. H. Zhao and J. Ruan, "Dynamic Analysis of High-Order Cohen-Grossberg Neural Networks with Time Delay,"

Chaos, Solitons & Fractals, Vol. 32, No. 4, 2007, pp. 1538-1546.

15. Hassan A. Zedan, "Solution of (3 + 1) - Dimensional Nonlinear Cubic Schrodinger Equation by Differential Transform Method," Mathematical Problems in Engineering, Vol. 2012, 2012, 14 p.

CHAPTER 5

Boundedness and Oscillation of Third Order Neutral Differential Equations with Deviating Arguments

Elmetwally M. Elabbasy, Magdy Y. Barsoum, Osama Moaaz

Department of Mathematics, Faculty of Science,
Mansoura University, Mansoura, Egypt

ABSTRACT

We consider the third-order neutral functional differential equations with deviating argu- ments. A new theorem is presented that improves a number of results reported in the literature. Examples are included to illustrate new results.

KEYWORDS

Oscillation, Third Order, Neutral Delay Differential Equations

1. INTRODUCTION

In this paper we consider third order neutral differential equations of the form

$$\left(r_2(t)\left(r_1(t)z'(t)\right)'\right)' + \int_\alpha^\beta q(t,\xi) f\left(x\left(g(t,\xi)\right)\right) \mathrm{d}\sigma(\xi) = 0, t \geq t_0, \tag{1}$$

where $z(t) = x(t) \pm p(t)x(\tau(t))$ and the following conditions are satisfied

(A$_1$) $p \in C(I,\mathbb{R}), 0 < p(t) \leq p < 1$ and $I = [t_0, \infty)$,

(A$_2$) $\tau \in C(I,\mathbb{R}), \tau(t) < t$, τ is strictly increasing, $\lim_{t\to\infty} \tau(t) = \infty$ and we define $\tau^0(t) = t, \tau^j(t) = \tau(\tau^{j-1}(t)), j = 1,2,\cdots,$

(A$_3$) $r_i \in C(I,(0,\infty)), \int_{t_0}^\infty r_i^{-1}(t)\mathrm{d}t = \infty, i = 1,2,$

(A$_4$) $f \in C(\mathbb{R},\mathbb{R})$, f is non-decreasing and $uf(u) > 0$ for $u \neq 0$,

(A$_5$) $q \in C(I \times [\alpha,\beta],[0,\infty)]$, and $q(t,\xi)$ is not zero on any half line $[t_\mu,\infty) \times [\alpha,\beta], t_\mu \geq t_0$,

(A$_6$) $g \in C(I \times [\alpha,\beta],\mathbb{R})$, $g(t,\xi) \leq t$ for $t \geq t_0$ and $\xi \in [\alpha,\beta]$, $g(t,\xi)$ is continuous, has positive partial derivative on $I \times [\alpha,\beta]$ with respect to t, nondecreasing with respect to ξ and $\lim_{t\to\infty} g(t,\xi) = \infty$,

(A$_7$) $\sigma \in C([\alpha,\beta],\mathbb{R})$, σ is nondecreasing and the integral of Equation (1) is in the sense Riemann-stieltijes.

We mean by a solution of Equation (1) a function $x(t):[t_x,\infty)\to\mathbb{R}$, $t_x \geq t_0$ such that $z(t)$, $r_1(t)z'(t)$, $r_2(t)(r_1(t)z'(t))'$ and $\left(r_2(t)(r_1(t)z'(t))'\right)'$ exist and are continuous on $[t_x,\infty)$. A nontrivial solution of (1) is called oscillatory if it has arbitrarily large zeros, otherwise it is called non-oscillatory.

Asymptotic properties of solutions of differential equations of the second and third order have been subject of intensive studying in the literature. This problem for neutral differential equations has received considerable attention in recent years (see [1] - [11]).

Recently, in [12] by using Riccati technique, have established some general oscillation criteria for third-order neutral differential equation

$$\left(r_2(t)\left(r_1(t)\left(x(t)+px(t-\tau)\right)'\right)'\right)'+q(t)x(t-\sigma)=0.$$

In [3] , Candan presented several oscillation criteria for third order neutral delay differential equation

$$\left(r_2(t)\left(r_1(t)\left(x(t)+px(t-\tau)\right)'\right)'\right)'+\int_a^\beta q(t,\xi)x(g(t,\xi))\mathrm{d}\sigma(\xi)=0.$$

[9] and [13] obtained some oscillation criteria for study third order nonlinear neutral differential equations

$$\left(r_2(t)\left(r_1(t)\left(x(t)+px(t-\tau)\right)'\right)'\right)'+q(t)f\left(x(t-\sigma)\right)=0$$

and

$$\left(r_2(t)\left(r_1(t)\left(x(t)-px(\tau(t))\right)'\right)'\right)'+q(t)f\left(x(g(t))\right)=0.$$

In this paper, we establish some oscillation criteria for Equation (1), which complement and extend the results in [3] [13] .

We begin with analyzing of the asymptotic behavior of possible non-oscillatory solutions of the Equation (1) in the case when $z(t)=x(t)-px(\tau(t))$. Let $x(t)$ be a non-oscillatory solution of (1) on $[t_0,\infty)$. From (1) it follows that the function $z(t)$ has to be eventually of constant sign, so either

(a) $x(t)z(t)>0$

or

(b) $x(t)z(t)<0$

for all sufficiently large t. Denote by N^+ [or N^-] the set of all non-oscillatory solutions $x(t)$ of the Equation (1) such that (a) [or (b)] is satisfied. We begin with some useful lemmas.

Lemma 1.1 Let $z(t)=x(t)-px(\tau(t))$. Assume that (A_1) and (A_2) hold and x be continuous non-oscillatory solution of the functional inequality (a). Then

$$\lim_{t\to\infty}x(t)=0 \text{ and } \lim_{t\to\infty}z(t)=0.$$

Lemma 1.2 Let $z(t)=x(t)-px(\tau(t))$. Assume that (A_1) and (A_2) hold and x be continuous non-oscillatory solution of the functional inequality (b). If $\lim_{t\to\infty}z(t)=0,$ then

$\lim_{t\to\infty} x(t) = 0.$

These lemmas are modifications of the Lemma 1 in the paper [14] and the Lemma 2 in the paper [13].

2. MAIN RESULTS

In this part, for the sake of convenience, we introduce the following notation:

$$\Theta(t) = \int_\alpha^\beta q(t,\xi)\mathrm{d}\sigma(\xi) \text{ and } R(t) = \int_{t_0}^t r_2^{-1}(s)\mathrm{d}s.$$

2.1. Oscillation Criteria If $z(t) = x(t) - px(\tau(t))$

In this section, we will establish some oscillation criteria for Equation (1) in the case when $z(t) = x(t) - px(\tau(t))$ and $p(t) = p$.

Lemma 2.1 Let x be a bounded positive solution of Equation (1) on the interval I. Then there exists a $T_0 \geq t_0$ such that $z(t)$ has the following properties:

$$z'(t) < 0, (r_1(t)z'(t))' > 0 \text{ and } \left(r_2(t)(r_1(t)z'(t))'\right)' \leq 0 \text{ for } t \geq T_0 \tag{2}$$

Proof. Let x be a bounded positive solution of Equation (1) on the interval I. From (A_1), (A_2) and (A_6), there exists a $t_1 \geq t_0$ such that $x(t) > 0$, $x(\tau(t)) > 0$ and $x(g(t,\xi)) > 0$ for $t \geq t_1$. Then $z(t)$ is bounded and non-oscillatory. Thus, Equation (1) implies that

$$\left(r_2(t)\left(r_1(t)z'(t)\right)'\right)' = -\int_\alpha^\beta q(t,\xi) f\left(x\left(g(t,\xi)\right)\right) d\sigma(\xi) \le 0.$$

Hence, the function $r_2(t)\left(r_1(t)z'(t)\right)'$ is a non-increasing and of one sign. We claim that $r_2(t)\left(r_1(t)z'(t)\right)' > 0$ for $t \ge t_1$. Suppose that $r_2(t)\left(r_1(t)z'(t)\right)' < 0$ for $t \ge t_2 \ge t_1$. Then there exists a $t_3 \ge t_2$ and constant $K_1 > 0$ such that

$$r_2(t)\left(r_1(t)z'(t)\right)' < -K_1 \quad \text{for } t \ge t_3.$$

By integrating the last inequality from t_3 to t, we get

$$r_1(t)z'(t) < r_1(t_3)z'(t_3) - K_1 \int_{t_3}^t \frac{1}{r_2(s)} ds.$$

Letting $t \to \infty$, from (A3), we have $\lim_{t\to\infty} r_1(t)z'(t) = -\infty$. Then there exists a $t_4 \ge t_3$ and constant $K_2 > 0$ such that

$$r_1(t)z'(t) < -K_2 \quad \text{for } t \ge t_4$$

By integrating this inequality from t_4 to t and using (A3), we get $\lim_{t\to\infty} z(t) = -\infty$. This yields that $x \in N^-$ and this contradicts the Lemma 1.1. Now we have $r_2(t)\left(r_1(t)z'(t)\right)' > 0$ for $t \ge t_2$. Hence $r_1(t)z'(t)$ is increasing function and we have two possible cases for $r_1(t)z'(t)$ either $r_1(t)z'(t) > 0$ eventually or $r_1(t)z'(t) < 0$ eventually for $t \ge t_3 \ge t_2$. If $r_1(t)z'(t) > 0$ for $t \ge t_3$, then there exist a $t_4 \ge t_3$ and a constant $L > 0$ such that

$r_1(t)z'(t) > L$ for $t \geq t_4$

By integrating this inequality from t_4 to t and using (A3), we get $\lim_{t\to\infty} z(t) = \infty$. This means that $z(t) > 0$ and we get $x(t) \geq z(t)$ for all sufficiently large t. Then $\lim_{t\to\infty} z(t) = \infty$, which contradicts the boundedness of $x(t)$. Hence, $z'(t) < 0$ for $t \geq t_3$.

Theorem 2.1 if

$$\limsup_{t\to\infty} \int_{g(t,\beta)}^{t} \Theta(s) \int_{g(s,\beta)}^{g(t,\beta)} \frac{1}{r_1(u)} \int_{u}^{g(t,\beta)} \frac{1}{r_2(v)} dv du ds > 0. \tag{3}$$

Then every bounded solution $x(t)$ of Equation (1) is either oscillatory or tends to zero.

Proof. Let x be a bounded non-oscillatory solution of Equation (1) on the interval I. Without loss of generality we may assume that $x(t) > 0$. From Lemma 2.1, we get that (2) holds. New, we have

$$x(t) = z(t) + px(\tau(t)) = z(t) + pz(\tau(t)) + p^2 x(\tau^2(t)),$$

for all sufficiently large t. Repeating this procedure and the monotonicity of z, we obtain that there exists an integer $n \geq 0$ such that $\tau^{n+1}(g(t,\xi)) \geq T_0$ and

$$x(g(t,\xi)) = \sum_{r=0}^{n} p^r z(\tau^r(g(t,\xi))) + p^{n+1} x(\tau^{n+1}(g(t,\xi))) \geq \lambda z(g(t,\xi)),$$

where $\lambda = \sum_{r=0}^{n} p^r = \frac{1-p^{n+1}}{1-p} > 1$. Hence, we get

$$f\left(x\left(g(t,\xi)\right)\right)\geq f\left(\lambda z\left(g(t,\xi)\right)\right)\geq f\left(z\left(g(t,\xi)\right)\right)\geq f\left(z\left(g(t,\beta)\right)\right). \tag{4}$$

Thus, from Equation (1), we obtain

$$\left(r_2(t)\left(r_1(t)z'(t)\right)'\right)'\leq -f\left(z\left(g(t,\beta)\right)\right)\Theta(t). \tag{5}$$

Now, since $z(t)$ is bounded decreasing function, then there exist γ such that

$$\lim_{t\to\infty} z(t)=\gamma.$$

If $z(t)<0$ for $t\geq T_1\geq T_0$, then $\gamma<0$ and which contradicts the Lemma 1.1. Therefore $z(t)>0$ for $t\geq T_1$ and $\gamma\geq 0$. We shall prove that $\gamma=0$. Let $\gamma>0$. For $t>u>T\geq T_1$, we obtain

$$\int_u^{g(t,\beta)}\frac{1}{r_2(v)}\left[r_2(v)\left(r_1(v)z'(v)\right)'\right]\mathrm{d}v=r_1\left(g(t,\beta)\right)z'\left(g(t,\beta)\right)-r_1(u)z'(u).$$

Thus, form Lemma 2.1, we get

$$\begin{aligned} r_1(u)z'(u) &\leq -\int_u^{g(t,\beta)}\frac{1}{r_2(v)}\left[r_2(v)\left(r_1(v)z'(v)\right)'\right]\mathrm{d}v \\ &\leq -\left[r_2\left(g(t,\beta)\right)\left(r_1\left(g(t,\beta)\right)z'\left(g(t,\beta)\right)\right)'\right]\int_u^{g(t,\beta)}\frac{1}{r_2(v)}\mathrm{d}v. \end{aligned} \tag{6}$$

So, for $t>u>s>T$, we have

$$z\left(g(t,\beta)\right)-z\left(g(s,\beta)\right)=\int_{g(s,\beta)}^{g(t,\beta)}\frac{1}{r_1(u)}\left[r_1(u)z'(u)\right]du.$$

Hence, from (6), we get

$$z(g(t,\beta)) - z(g(s,\beta)) \le -G(g(t,\beta)) \int_{g(s,\beta)}^{g(t,\beta)} \frac{1}{r_1(u)} \int_u^{g(t,\beta)} \frac{1}{r_2(v)} \mathrm{d}v\mathrm{d}u \tag{7}$$

where $G(y) = r_2(y)(r_1(y)z'(y))'$. Let us define function

$$F(s,t) = (G(s) - G(g(t,\beta))) \int_s^t \frac{z'(g(u,\beta))g'(u,\beta)}{f(z(g(u,\beta)))} \mathrm{d}u, \quad t \ge s \ge T$$

We note that $F(t,t) = 0 = F(g(t,\beta),t)$. Deriving $F(s,t)$ partially with respect to s and using Lemma 2.1, (A_4) and (A_6), we get

$$F_s'(s,t) \ge G'(s) \int_s^t \frac{z'(g(u,\beta))g'(u,\beta)}{f(z(g(u,\beta)))} \mathrm{d}u + G(g(t,\beta)) \frac{z'(g(s,\beta))g'(s,\beta)}{f(z(g(s,\beta)))}$$

From (5), we have $G'(s) \le -f(z(g(s,\beta)))\Theta(s)$. Hence, we obtain

$$F_s'(s,t) \ge -\Theta(s) f(z(g(s,\beta))) \int_s^t \frac{z'(g(u,\beta))g'(u,\beta)}{f(z(g(u,\beta)))} \mathrm{d}u + G(g(t,\beta)) \frac{z'(g(s,\beta))g'(s,\beta)}{f(z(g(s,\beta)))} \tag{8}$$

By (A_4) and (A_6), we get

$$\int_s^t \frac{z'(g(u,\beta))g'(u,\beta)}{f(z(g(u,\beta)))} \mathrm{d}u \le \frac{1}{f(z(g(s,\beta)))} \int_s^t \mathrm{d}z(g(u,\beta)) = \frac{1}{f(z(g(s,\beta)))} (z(g(t,\beta)) - z(g(s,\beta))) \le 0$$

Thus, from (7), we have

$$\int_s^t \frac{z'(g(u,\beta))g'(u,\beta)}{f(z(g(u,\beta)))}du \le \frac{-G(g(t,\beta))}{f(z(g(s,\beta)))}\int_{g(s,\beta)}^{g(t,\beta)}\frac{1}{r_1(u)}\int_u^{g(t,\beta)}\frac{1}{r_2(v)}dvdu. \tag{9}$$

Then, substituting (8) in (9), it follows that

$$F_s'(s,t) \ge G(g(t,\beta))\left[\Theta(s)\int_{g(s,\beta)}^{g(t,\beta)}\frac{1}{r_1(u)}\int_u^{g(t,\beta)}\frac{1}{r_2(v)}dvdu + \frac{z'(g(s,\beta))g'(s,\beta)}{f(z(g(s,\beta)))}\right].$$

By integrating this inequality from $g(t,\beta)$ to t with respect to s, we obtain

$$0 \ge \int_{g(t,\beta)}^t \Theta(s)\int_{g(s,\beta)}^{g(t,\beta)}\frac{1}{r_1(u)}\int_u^{g(t,\beta)}\frac{1}{r_2(v)}dvduds + \int_{z(g(g(t,\beta),\beta))}^{z(g(t,\beta))}\frac{d\omega}{f(\omega)}. \tag{10}$$

where $G(g(t,\beta))>0$. Since $\lim_{t\to\infty} z(t)=\gamma>0$, we get

$$\lim_{t\to\infty}\int_{z(g(g(t,\beta),\beta))}^{z(g(t,\beta))}\frac{d\omega}{f(\omega)}=0$$

Hence, from (10), we have

$$\limsup_{t\to\infty}\int_{g(t,\beta)}^t \Theta(s)\int_{g(s,\beta)}^{g(t,\beta)}\frac{1}{r_1(u)}\int_u^{g(t,\beta)}\frac{1}{r_2(v)}dvduds \le 0,$$

which contradicts (3). Therefore, $\lim_{t\to\infty} z(t)=0$ and according to the Lemma 1.2 we have that $\lim_{t\to\infty} x(t)=0$.

In the following Theorem, we establish some sufficient conditions for boundedness and oscillation of Equation (1) under the condition

$$\liminf_{u\to 0}\frac{f(u)}{u}>0 \tag{11}$$

Theorem 2.2 Let (11) holds. If there exist an integer $n\geq 0$ such that

$$\limsup_{t\to\infty}\int_{g(t,\beta)}^{t}\Theta(s)\int_{g(s,\beta)}^{g(t,\beta)}\frac{1}{r_1(u)}\int_{u}^{g(t,\beta)}\frac{1}{r_2(v)}\,dvduds>\frac{1-p}{1-p^{n+1}}\limsup_{u\to 0}\frac{u}{f(u)}, \tag{12}$$

then every bounded solution $x(t)$ of Equation (1) is oscillatory.

Proof. Let x be a bounded non-oscillatory solution of Equation (1) on the interval I. Without loss of generality we may assume that $x(t)>0$. We can proceed exactly as in the proof of Theorem 2.1 and we use the fact that (12) implies (3). Hence, we get a non-oscillatory solution with the properties $x(t)>0$, $z(t)>0$, $z'(t)<0$ and $(r_1(t)z'(t))'>0$ for $t\geq T_0$, $\lim_{t\to\infty}z(t)=0$ and $\lim_{t\to\infty}x(t)=0$. New, from (4), there exists $T_1\geq T_0$ such that

$$f(x(g(t,\xi)))\geq f(\lambda z(g(t,\xi)))\geq f(\lambda z(g(t,\beta)))\text{ for }t\geq T_1.$$

Thus, Equation (1) implies that

$$\left(r_2(t)(r_1(t)z'(t))'\right)'\leq -f(\lambda z(g(t,\beta)))\Theta(t).$$

By integrating this inequality from $g(t,\beta)$ to t, we get

$$G(t)\leq G(g(t,\beta))-\int_{g(t,\beta)}^{t}f(\lambda z(g(s,\beta)))\Theta(s)ds.$$

where $G(y)=r_2(y)\left(r_1(y)z'(y)\right)'>0$. Thus, we obtain

$$0\le G\left(g(t,\beta)\right)-M\lambda\int_{g(t,\beta)}^{t}z\left(g(s,\beta)\right)\Theta(s)\,ds. \tag{13}$$

where $M=\inf\left\{\frac{f(u)}{u}:0\le|u|\le\left|\lambda z\left(g(t,\beta)\right)\right|\right\}$. Since $z(t)>0$, from the Inequality (7), we get

$$-z\left(g(s,\beta)\right)\le -G\left(g(t,\beta)\right)\int_{g(s,\beta)}^{g(t,\beta)}\frac{1}{r_1(u)}\int_{u}^{g(t,\beta)}\frac{1}{r_2(v)}\,dv\,du. \tag{14}$$

Combining (13) and (14), we have

$$0<G\left(g(t,\beta)\right)\left(1-M\lambda\int_{g(t,\beta)}^{t}\Theta(s)\int_{g(s,\beta)}^{g(t,\beta)}\frac{1}{r_1(u)}\int_{u}^{g(t,\beta)}\frac{1}{r_2(v)}\,dv\,du\,ds\right).$$

Hence, we get

$$\int_{g(t,\beta)}^{t}\Theta(s)\int_{g(s,\beta)}^{g(t,\beta)}\frac{1}{r_1(u)}\int_{u}^{g(t,\beta)}\frac{1}{r_2(v)}\,dv\,du\,ds<\frac{1}{M\lambda},$$

for $t\ge s\ge T_1$ and this contradicts the condition (12).

Corollary 2.1 Let (11) holds. If

$$\limsup_{t\to\infty}\int_{g(t,\beta)}^{t}\Theta(s)\int_{g(s,\beta)}^{g(t,\beta)}\frac{1}{r_1(u)}\int_{u}^{g(t,\beta)}\frac{1}{r_2(v)}\,dv\,du\,ds>(1-p)\limsup_{u\to 0}\frac{u}{f(u)} \tag{15}$$

then every bounded solution $x(t)$ of Equation (1) is oscillatory.

Example 2.1 Consider the differential equation

$$\left(\sqrt{t}\left(\sqrt{t}\left(x(t)-\frac{1}{2}x\left(\left(\sqrt{t}-1\right)^2\right)\right)'\right)'\right)'+\int_0^1\frac{3\xi^2}{\sqrt{t}}x\left(\left(\sqrt{t}-1\right)^2\xi\right)d\xi=0,$$

where $t>1$. We have

$$\limsup_{t\to\infty}\int_{g(t,\beta)}^{t}\Theta(s)\int_{g(s,\beta)}^{g(t,\beta)}\frac{1}{r_1(u)}\int_{u}^{g(t,\beta)}\frac{1}{r_2(v)}\mathrm{d}v\mathrm{d}u\mathrm{d}s=\frac{4}{3},$$

and $(1-p)\limsup_{u\to0}\frac{u}{f(u)}=\frac{1}{2}<\frac{4}{3}$. Thus, all conditions of Corollary 2.1 are satisfied then all bounded solutions of the above equation are oscillatory.

Remark 2.1 If $\alpha=0,\beta=1,q(t,\xi)=q(t),g(t,\xi)=g(t)$ and $\sigma(\xi)=\xi$ then, our results extend the results in [13].

2.2. Oscillation Criteria If $z(t)=x(t)-px(\tau(t))$

In this section, we will present some oscillation criteria for Equation (1) under the case $z(t)=x(t)+p(t)x(\tau(t))$ and the condition

$$\frac{f(u)}{u}\geq k>0 \text{ for } u\neq0 \text{ and } t\geq t_0 \tag{16}$$

Lemma 2.2 If $x(t)$ is an eventually positive solution of (1), then for sufficiently large t, there are only two possible cases:

(i) $z(t)>0,z'(t)<0$ and $(r_1(t)z'(t))'>0$

(ii) $z(t)>0, z'(t)>0$ and $\left(r_1(t)z'(t)\right)'>0.$

Proof. The proof of this lemma is similar to the proof Lemma 1 in [9] and we omit the details.

Theorem 2.3 Let (16) holds. If

$$\int_{t_0}^{\infty}\frac{1}{r_1(v)}\int_{v}^{\infty}\frac{1}{r_2(u)}\int_{u}^{\infty}\Theta(s)\,dsdudv=\infty, \tag{17}$$

and there exist a positive real function $\rho(t)$ such that

$$\limsup_{t\to\infty}\int_{t_0}^{t}\left(k\rho(s)(1-p)\Theta(s)-\frac{\rho^2(s)r_1(g(s,a))}{4\rho(s)R(g(s,a))g'(s,a)}\right)ds=\infty. \tag{18}$$

Then every solution of Equation (1) is either oscillatory or tends to zero.

Proof. Let x be a non-oscillatory solution of Equation (1) on the interval I. Without loss of generality we may assume that $x(t)>0$. Then there exists a $t_1\geq t_0$ such that $x(t)>0$, $x(\tau(t))>0$ and $x(g(t,\xi))>0$ for $t\geq t_1$. By Lemma 2.2, we have two cases for $z(t)$. In the Case (i), since $z(t)>0$ and $z'(t)<0$, we get $\lim_{t\to\infty}z(t)=\mu\geq 0$. Let $\mu>0$, then we have $\mu<z(t)<\mu+\varepsilon$ for all $\varepsilon>0$ and t enough large. Choosing $\varepsilon<\frac{1-p}{p}\mu$, we obtain

$$\begin{aligned}x(t)&=z(t)-p(t)x(\tau(t))\\&\geq\mu-pz(\tau(t))\geq L(\mu+\varepsilon)\geq Lz(t).\end{aligned}$$

where $L=\frac{\mu-p(\mu+\varepsilon)}{\mu+\varepsilon}>0$. Hence, from (1), (A6) and (16), we have

$$\left(r_2(t)\left(r_1(t)z'(t)\right)'\right)'\le -kL\int_\alpha^\beta q(t,\xi)z(g(t,\xi))\mathrm{d}\sigma(\xi)<-kL\mu\Theta(t).$$

By integrating two times from t to ∞, we get

$$-z'(t)>kL\mu\frac{1}{r_1(t)}\int_t^\infty\frac{1}{r_2(u)}\int_u^\infty\Theta(s)\mathrm{d}s\mathrm{d}u.$$

Integrating the last inequality from t_1 to ∞, we obtain

$$z(t_1)>kL\mu\int_{t_1}^\infty\frac{1}{r_1(v)}\int_v^\infty\frac{1}{r_2(u)}\int_u^\infty\Theta(s)\mathrm{d}s\mathrm{d}u\mathrm{d}v.$$

This contradicts to the condition (17), then $\lim_{t\to\infty}z(t)=0$, which implies that $\lim_{t\to\infty}x(t)=0$. In the Case (ii), since $z(t)>0$ and $z'(t)>0$. Then there exist a $t_1\ge t_0$ such that

$$x(t)\ge(1-p)z(t),$$

for $t\ge t_1$. Thus, from (1), (A4) and (A6), we get

$$\left(r_2(t)\left(r_1(t)z'(t)\right)'\right)'\le -k(1-p)z(g(t,a))\Theta(t). \tag{19}$$

Also, we have

$$r_1(t)z'(t)=r_1(t_1)z'(t_1)+\int_{t_1}^t\frac{r_2(s)\left(r_1(s)z'(s)\right)'}{r_2(s)}\mathrm{d}s\ge r_2(t)\left(r_1(t)z'(t)\right)'\int_{t_1}^t\frac{1}{r_2(s)}\mathrm{d}s,$$

Since $\left(r_2(t)\left(r_1(t)z'(t)\right)'\right)' \le 0$, we obtain

$$z'(g(t,a)) \ge \frac{r_2(t)\left(r_1(t)z'(t)\right)'}{r_1(g(t,a))} R(g(t,a)). \tag{20}$$

Now, we define

$$\omega(t) = \rho(t)\frac{r_2(t)\left(r_1(t)z'(t)\right)'}{z(g(t,a))}.$$

By differentiating and using (19) and (20), we get

$$\begin{aligned}\omega'(t) &\le -k\rho(t)(1-p)\Theta(t) + \frac{\rho'(t)}{\rho(t)}\omega(t) - \frac{R(g(t,a))g'(t,a)}{\rho(t)r_1(g(t,a))}\omega^2(t) \\ &= -k\rho(t)(1-p)\Theta(t) + \frac{\rho'^2(t)r_1(g(t,a))}{4\rho(t)R(g(t,a))g'(t,a)} \\ &\quad - \left(\sqrt{\frac{R(g(t,a))g'(t,a)}{\rho(t)r_1(g(t,a))}}\omega(t) - \rho'(t)\sqrt{\frac{r_1(g(t,a))}{4\rho(t)R(g(t,a))g'(t,a)}}\right)^2.\end{aligned}$$

Hence, we obtain

$$\omega'(t) \le -k\rho(t)(1-p)\Theta(t) + \frac{\rho^{2}(t)r_1(g(t,a))}{4\rho(t)R(g(t,a))g'(t,a)}.$$

By integrating the above inequality from t_1 to t we have

$$\omega(t) \le \omega(t_1) - \int_{t_1}^{t}\left(k\rho(s)(1-p)\Theta(s) - \frac{\rho'^2(s)r_1(g(s,a))}{4\rho(s)R(g(s,a))g'(s,a)}\right)ds.$$

Taking the superior limit as $t \to \infty$ and using (18), we get $\omega(t) \to -\infty$ which contradicts that $\omega(t) > 0$. This completes the proof of Theorem 2.3. □

Remark 2.2 We can rewrite the condition (17) in the Theorem 2.3 as following

$$\int_{t_0}^{\infty}\left(\int_{t_0}^{r}\frac{\mathrm{d}v}{r_1(v)}\left(\int_{v}^{r}\frac{\mathrm{d}u}{r_2(u)}\right)\right)\Theta(r)\,\mathrm{d}r=\infty.$$

Remark 2.3 If $\tau(t)=t-\tau$ and $f(x)=x$, then our results extend the results in [3] .

Example 2.2 Consider the differential equation

$$\left(\frac{1}{t}\left(\frac{1}{t}z'(t)\right)'\right)'+\int_0^1\frac{t^2e^{t^2\xi}}{e^{t^2}-1}x(\xi(t-1))\left(x^2(\xi(t-1))+2\right)\mathrm{d}\xi=0,$$

where $t>1$ and $z(t)=x(t)+\frac{1}{2}x(t-1)$. Choosing $\rho(t)=1$ and $k=2$. Thus, all conditions of Theorem 2.3 are satisfied then every solutions of this equation is either oscillatory or tends to zero.

REFERENCES

1. Agarwal, R.P., Aktas, M.F. and Tiryaki, A. (2009) On Oscillation Criteria for Third Order Nonlinear Delay Differential Equations. Archivum Mathematicum (Brno), 45, 1-18.

2. Aktas, M.F., Tiryaki, A. and Zafer, A. (2010) Oscillation Criteria for Third-Order Nonlinear Functional Differential Equations. Applied Mathematics Letters, 23, 756-762.

3. Candan, T. and Dahiya, R.S. (2005) Functional Differential Equations of Third Order. Electronic Journal of Differential Equations, Conference 12, 47-56.

4. Dzurina, J. and Kotorova, R. (2008) Asymptotic Properties of Trinomial Delay Differential Equations. Archivum Mathematicum (Brno), 44, 149-158.

5. Dzurina, J. and Kotorova, R. (2009) Properties of the Third Order Trinomial Differential Equations with Argument. Nonlinear Analysis, 71, 1995-2002.

6. Elabbasy, E.M., Hassan, T.S. and Moaaz, O. (2012) Oscillation Behavior of Second Order Nonlinear Neutral Differential Equations with Deviating Arguments. Opuscula Mathematica, 32, 719-730.

7. El-sheikh, M.M., Sallam, R. and Mohamady, N. (2013) On the Oscillation of Third Order Neutral Delay Differential Equations. Applied Mathematics & Information Sciences Letters, 1, 77-80.

8. Erbe, L.H., Kong, Q. and Zhang, B.G. (1994) Oscillation Theory for Functional Differential Equations. Marcel Dekker, Inc., New York.

9. Graef, J.R., Savithri, R. and Thandapani, E. (2002) Oscillatory Properties of Third Order Neutral Delay Differential Equations. Proceedings of the Fourth International Conference on Dynamical Systems and Differential Equations, Wilmington, 24-27 May 2002, 342-350.

10. Hale, J. (1977) Theory of Functional Differential Equations. 2nd Edition, Applied Mathematical Sciences Vol. 3, Springer, New York.

11. Su, M. and Xu, Z. (2012) Oscillation Criteria Of Certain Third Order Neutral Differential Equations. Differential Equations & Applications, 4, 221-232.

12. Li, T., Thandapani, E. and Graef, J.R. (2012) Oscillation of Third-Order Neutral Retarded Differential Equations. International Journal of Pure and Applied Mathematics, 75, 511-520.

13. Mihalikova, B. and Kostikova, E. (2009) Boundedness and Oscillation of Third Order Neutral Differential Equations, Tatra Mountains Mathematical Publications, 43, 137-144.

14. Laros, J. and Kusano, T. (1990) On a Class of First Order Nonlinear Functional Differential Equations of Neutral Type. Czechoslovak Mathematical Journal, 40, 475-490.

CHAPTER 6

Differential Transform Method for Some Delay Differential Equations

Baoqing Liu[1], Xiaojian Zhou[2], Qikui Du[3]

[1] School of Applied Mathematics, Jiangsu Provincial Key Laboratory for NSLSCS, Nanjing University of Finance and Economics, Nanjing, China

[2] School of Science, Jiangsu Provincial Key Laboratory for NSLSCS, Nantong University, Nantong, China

[3] School of Mathematical Sciences, Jiangsu Key Laboratory for NSLSCS, Nanjing Normal University, Nanjing, China

ABSTRACT

This paper concentrates on the differential transform method (DTM) to solve some delay differential equations (DDEs). Based on the method of steps for DDEs and using the computer algebra system Mathematica, we successfully apply DTM to find the analytic solution to some DDEs, including a neural delay differential equation. The results confirm the feasibility and efficiency of DTM.

KEYWORDS

Differential Transform Method, Delay Differential Equation, Method of Steps, Analytic Solution, Approximate Solution

1. INTRODUCTION

The differential transform method (DTM) is a semi analytical-numerical technique depending on Taylor series for solving integral-differential equations (IDEs). The method was first introduced by Pukhov [1] for solving linear and nonlinear initial value problems in physical processes. Zhou, at the same time, had also introduced DTM to study electrical circuits [2] . Since the main advantage of this method is that it can be applied directly to nonlinear ordinary and partial differential equations without requiring linearization, discretization or perturba- tion, it has been studied and applied during the last two decades widely. DTM has been used to obtain numerical and analytical solutions of ordinary differential equations [3] , partial differential equations [4] , eigenvalue pro- blems [5] , differential algebraic equations [6] [7] , integral equations [8] and so on.

Delay differential equations (DDEs) arise in many applied fields, such as control technology, communication networks, and biological population management, and hence they have attracted considerable attention. There are many papers devoted to the problem of approximate solution of DDEs [9] - [15] . Recently, F. Karako and H. Bereketoğlu [13] extend the method of differential transformation for solving the following two types of DDEs:

$$\begin{cases} y'(t) = f\left(t, y(t), y(at)\right), & 0 \le t \le T, \\ y(0) = y_0, \end{cases} \tag{1}$$

and

$$\begin{cases} y'(t) = f\left(t, y(t), y(t-\tau)\right), & 0 \le t \le T, \\ y(0) = y_0, \end{cases} \tag{2}$$

with $0 < a < 1$ and $\tau > 0$ and the constant $T > 0$.

It should be pointed out that the solution to DDEs (2) maybe be non-unique (see Section 2 in [16]). So usually, researchers pay more attention to the following DDEs, instead of (2)

$$\begin{cases} y'(t) = f\left(t, y(t), y(t-\tau)\right), & 0 \le t \le T, \\ y(t) = g(t), & t \le 0, \end{cases} \tag{3}$$

where $g(t)$ is a given function, called initial function.

In this paper, we will apply DTM to find the analytic solution to DDEs (3) with the help of the computer algebra system Mathematica. Thus, in some sense, our work can be viewed as a supplement to [13].

2. DIFFERENTIAL TRANSFORM

The basic theory of differential transform can be found in [1] [2] , in this section we will state it in brief.

Consider a function $y(t)$ be analytic in the time domain I, and let $t_0 \in I$. The function $y(t)$ is then represented by one series whose center is located at t_0. The differential transform of the function $y(t)$ is the form

$$Y(k) = \frac{1}{k!} \frac{\mathrm{d}^k y(t)}{\mathrm{d}t^k}\bigg|_{t=t_0} \tag{4}$$

where $Y(k)$ is the transformed function of the original function $y(t)$.

Differential inverse transformation of $Y(k)$ is defined as follows:

$$y(t) = \sum_{k=0}^{\infty} Y(k)(t - t_0)^k \tag{5}$$

From (4) and (5), it is easy to see that the concept of the differential transformation is derived from the Taylor series expansion. By our assumption, t_0 is taken as zero, then the function $y(t)$ is expressed by a finite series and (5) can be written as

$$y(t) = \sum_{k=0}^{\infty} Y(k) t^k$$

In this study, we use the lower case letters to represent the original functions and upper case letters to stand for the transformed functions (T-functions). The fundamental mathematical operations performed by differential transform method are listed in Table 1.

3. DTM FOR DDES (3)

There are many methods to deal with the delay differential Equation (3). For example, linear multistep (LM) methods, Runge-Kutta (RK) methods, waveform relaxation (WR) methods, etc. However, the basic idea to solve the DDE (3) is to solve the following system of ODEs step by step:

$$y_i'(t) = f\left(t, y_i(t), y_{i-1}(t-\tau)\right), \quad i = 1,2,\cdots, \quad t \in \left[(i-1)\tau, i\tau\right] \tag{6}$$

with $y_0(t) = g(t)$. In brief, this idea is to shift the interval from $[(i-1)\tau, i\tau]$ to $[i\tau, (i+1)\tau]$ and extend the solution from $[0, i\tau]$ to $[0, (i+1)\tau]$ by using the component in the current interval. This procedure can, in principle, be continued as far as desired. It is called, quite naturally, the method of steps [16] .

Using the basic idea of the method of steps, first, we apply the DTM to find the solution to the following ODEs:

$$y_1'(t) = f\left(t, y_1(t), g(t-\tau)\right)$$

with $t \in [0, \tau]$.

Suppose the approximate solution is given by

$$y_1(t) = \sum_{k=0}^{\infty} a_{1k} t^k, \quad t \in [0, \tau]$$

If $\tau > T$, $y_1(t)$ is the solution to (3). Otherwise, we should continue to find the solution in the interval $[\tau, 2\tau]$. At this time, we should solve the following ODEs

$$y_2'(t) = f\left(t, y_2(t), y_1(t-\tau)\right), \quad t \in [\tau, 2\tau]$$

Table 1. Estimates of τ_0, λ_0, and λ_1 from simulated data.

λ_0	λ_1	Sample Size	MLE			LSE		
			τ_0	λ_0	λ_1	τ_0	λ_0	λ_1
0.3	0.1	100	2.8	0.33	0.150	3.925	0.239	0.159
		200	2.701	0.315	0.156	5.117	0.233	0.157
		300	2.979	0.312	0.147	5.917	0.222	0.155
0.25	0.15	100	2.809	0.271	0.173	3.860	0.234	0.188
		200	2.93	0.263	0.176	3.808	0.254	0.184
		300	3.146	0.262	0.171	4.232	0.251	0.182
0.2	0.15	100	3.44	0.208	0.161	4.136	0.212	0.169
		200	3.403	0.208	0.159	4.72	0.225	0.166
		300	3.261	0.208	0.158	5.111	0.242	0.164

Applying the DTM to the differential equation above again, we will obtain the following solution

$$y_2(t)=\sum_{k=0}^{\infty}a_{2k}(t-\tau)^k,\quad t\in[\tau,2\tau]$$

Of course, we should go on if $2\tau<T$ holds also. In generally, applying the DTM to ODEs (6), we can obtain the analytic solution

$$y_{n+1}(t)=\sum_{k=0}^{\infty}a_{nk}(t-n\tau)^k,\quad t\in[n\tau,(n+1)\tau]$$

until for some n, $n\tau<T\le(n+1)\tau$. In fact, after necessary steps, we have the following solution to (3)

$$y(t)=\begin{cases}y_1(t), & t\in[0,\tau],\\ y_2(t), & t\in[\tau,2\tau],\\ \vdots & \\ y_{n+1}(t), & t\in[n\tau,T].\end{cases}$$

Remark 1 If $y_1(t)=g(t)$, we can conclude that $y=g(t)$ is the analytic solution to (3) directly.

Remark 2 If $y_{i+1}(t)=y_i(t)$ for some integer i, we can conclude that the analytic solution (3) is

$$y(t)=\begin{cases}y_1(t), & t\in[0,\tau],\\ \vdots & \\ y_{i-1}(t), & t\in[(i-2)\tau,(i-1)\tau],\\ y_i(t), & t\in[(i-1)\tau,T].\end{cases}$$

Remark 3 If we want to improve the accuracy of the approximate solution in each interval, we can combine the above method with the multi-step method given by [17] .

Remark 4 In fact, the DTM based on the method of steps can also be applied to solve the following neutral delay differential equations

$$\begin{cases} y'(t) = f\left(t, y(t), y(t-\tau), y'(t-\tau)\right), & 0 \le t \le T, \\ y(t) = g(t), & t \le 0. \end{cases}$$

4. NUMERICAL EXPERIMENTS

In this section, four examples are given to show the performance of the DTM based on the method of steps. First, we want to solve the following simple but classical DDE to further illustrate the process of DTM.

Example 4.1 Consider the DDE [18]

$$\begin{cases} y'(t) = -y(t-1), & t \ge 0, \\ y(t) = 1, & -1 \le t \le 0. \end{cases} \tag{7}$$

First, since $\tau = 1$, we apply the DTM to obtain the solution in the interval $[0,1]$. In this interval, (7) can be written as $y'(t) = -1$, and the initial condition is $y(0) = 1$. Taking the differential transform, we have

$$(k+1)Y(k+1) = -\delta(k), \quad Y(0) = 1$$

It is easy to get

$$Y(k)=\begin{cases}1, & k=0,\\ -1, & k=1,\\ 0, & k\geq 2.\end{cases}$$

Thus we have the analytic solution $y(t)=1-t$ of (7) defined on $[0,1]$.

Second, we should continue to solve the following DDE:

$$\begin{cases}y'(t)=-y(t-1), & 1\leq t\leq 2,\\ y(t)=1-t, & 0\leq t\leq 1,\end{cases}$$

or equivalently,

$$y'(t)=t-2=-1+(t-1),\quad 1\leq t\leq 2 \tag{8}$$

with initial condition $y(1)=0$.

From (8), we have the following differential transform

$$(k+1)Y(k+1)=-\delta(k)+\delta(k-1),\quad Y(0)=0$$

It is easy to get

$$Y(k)=\begin{cases}0, & k=0,\\ -1, & k=1,\\ \dfrac{1}{2}, & k=2,\\ 0, & k\geq 3.\end{cases}$$

Thus we have the analytic solution

$$y(t)=-(t-1)+\frac{(t-1)^2}{2}=1-t+\frac{(t-1)^2}{2},\quad 1\le t\le 2$$

Now, if we want to obtain the solution in the interval [2, 3], we should deal with the following DDE:

$$\begin{cases} y'(t)=-y(t-1), & 2\le t\le 3, \\ y(t)=1-t+\dfrac{(t-1)^2}{2}, & 1\le t\le 2. \end{cases}$$

or equivalently,

$$y'(t)=(t-2)-\frac{(t-2)^2}{2},\quad 2\le t\le 3$$

with the initial condition $y(2)=-\frac{1}{2}$.

Then we have the following differential transform

$$(k+1)Y(k+1)=\delta(k-1)-\frac{1}{2}\delta(k-2),\quad Y(0)=-\frac{1}{2}$$

and get

$$Y(k)=\begin{cases} -\dfrac{1}{2}, & k=0, \\ 0, & k=1, \\ \dfrac{1}{2}, & k=2, \\ -\dfrac{1}{6}, & k=3, \\ 0, & k\ge 4. \end{cases}$$

Thus the analytic solution defined on [2, 3] is given by

$$y(t)=-\frac{1}{2}+\frac{(t-2)^2}{2}-\frac{(t-2)^3}{6}=1-t+\frac{(t-1)^2}{2}-\frac{(t-2)^3}{6},\quad 2\le t\le 3$$

The DTM can be proceed till the desire solution is obtained.

Example 4.2 Consider the nonlinear DDE of third-order [11] [13]

$$\begin{aligned} & y'''(x) = -y(x) - y(x-0.3) + e^{-x+0.3}, \qquad 0 \le x \le 1, \\ & y(0)=1,\ f'(0)=-1,\ y''(0)=1,\ y(x)=e^{-x}, \quad x \le 0. \end{aligned} \tag{9}$$

Since $\tau = 0.3$, according to the foregoing, we have the following ODE, defined in the interval [0, 0.3]

$$y_1'''(x) = -y_1(x)$$

Thus, applying DTM to the equation above, we obtain

$$(k+1)(k+2)(k+3)Y(k+3) = -Y(k)$$

The initial conditions lead to $Y(0)=1$, $Y(1)=-1$, and $Y(2)=\frac{1}{2}$. It is easy to have $Y(k)=\frac{(-1)^k}{k!}\ (k \ge 0)$. Thus we have the solution

$$y_1(x) = \sum_{k=0}^{\infty} \frac{(-1)^k}{k!} x^k = e^{-x}, \quad x \in [0, 0.3]$$

Noting that $y_1(x) = g(x) = e^{-x}$, Remark 1 tells us $y(x) = e^{-x}$ is the analytic solution to (9) in the whole interval $[0,1]$.

Remark 5 In [13], F. Karako and H. Bereketoğlu also apply DTM to DDE (9) where the initial function $y(x) = e^{-x}$, $x \le 0$ is omitted, i.e.

$$\begin{aligned} & y'''(x) = -y(x) - y(x-0.3) + e^{-x+0.3}, \quad 0 \le x \le 1, \\ & y(0)=1,\ f'(0)=-1,\ y''(0)=1, \qquad x \le 0. \end{aligned} \tag{10}$$

It's worth pointing out that, using the method given in [13] , only approximate solution can be obtained. On the other hand, for DDEs (3), the initial function has the vital role. Without it, the DDEs may have un-unique solution. In fact, Example 4.2 shows that $y(x)=e^{-x}$ is a solution to (10). Let's suppose $y(x)=e^{-x}+f(x)$ is

also a solution to (10), then $f(x)$ should satisfy the following DDEs

$$\begin{aligned} &f'''(x)=-f(x)-f(x-0.3), \quad 0\le x\le 1, \\ &f(0)=f'(0)=f''(0)=0, \qquad x\le 0. \end{aligned} \tag{11}$$

Figure 1 shows the solution to (11), together with it's first and second derivative value on $[0,1]$, with the initial function $g(x)=x^3$, which satisfies the initial conditions in (10) obviously. It can be seen $f(x)\neq 0$, then $y(x)=e^{-x}+f(x)$ is also the solution to (11). So, (11) has infinite solutions. Maybe, the authors "happen to" get the approximate solution of $y(x)=e^{-x}$.

Example 4.3 Consider a single delay equation with a stiffness parameter [10]

$$y'(t)=Ay(t)+y\left(t-\frac{3\pi}{2}\right)-A\sin(t) \tag{12}$$

with $y(0)=1$, $y(t)=e^{pt}+\sin(t)$, $t\in\left[-\frac{3\pi}{2},0\right]$.

Similarly, we should solve the following DDE limited in the interval $\left[0,\frac{3\pi}{2}\right]$:

$$y'(t)=Ay(t)+\mathrm{e}^{-\frac{3p\pi}{2}}\mathrm{e}^{pt}+\cos(t)-A\sin(t)$$

Thus applying DTM to this equation, we obtain

$$(k+1)Y(k+1)=AY(k)+\mathrm{e}^{-\frac{3p\pi}{2}}\frac{p^k}{k!}+\frac{1}{k!}\cos\left(\frac{\pi k}{2}\right)-A\frac{1}{k!}\sin\left(\frac{\pi k}{2}\right)$$

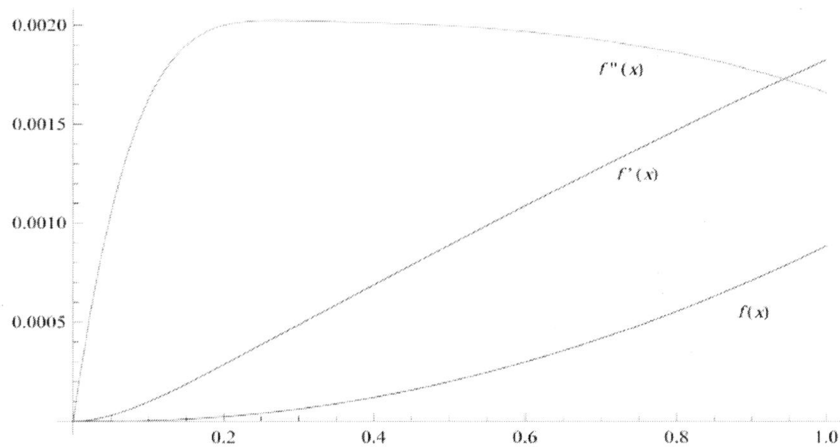

Figure 1. $f(x), f'(x)$ and $f''(x)$ of (11) with $g(x)=x^3$.

The initial values lead to $Y(0)=1$. With the help of Mathematica, we have

$$Y(1)=1+A+e^{-\frac{3p\pi}{2}}=\frac{1}{1!}+\frac{A}{1!}+\frac{1}{1!}e^{-\frac{3p\pi}{2}}\frac{A-p}{A-p},$$

$$Y(2)=\frac{1}{2}e^{-\frac{3p\pi}{2}}\left(A+A^2e^{\frac{3p\pi}{2}}+p\right)=\frac{1}{2!}A^2+\frac{1}{2!}e^{-\frac{3p\pi}{2}}(A+p)$$

$$=0+\frac{A^2}{2!}+\frac{1}{2!}e^{-\frac{3p\pi}{2}}\frac{A^2-p^2}{A-p},$$

$$Y(3)=\frac{1}{6}e^{-\frac{3p\pi}{2}}\left(A^2-e^{\frac{3p\pi}{2}}+A^3e^{\frac{3p\pi}{2}}+Ap+p^2\right)$$

$$=-\frac{1}{3!}+\frac{A^3}{3!}+\frac{1}{3!}e^{-\frac{3p\pi}{2}}\frac{A^3-p^3}{A-p},$$

$$Y(4)=\frac{1}{24}e^{-\frac{3p\pi}{2}}\left(A^3+A^4e^{\frac{3p\pi}{2}}+A^2p+Ap^2+p^3\right)$$

$$=0+\frac{A^4}{4!}+\frac{1}{4!}e^{-\frac{3p\pi}{2}}\frac{A^4-p^4}{A-p},$$

$$Y(5)=\frac{1}{120}e^{-\frac{3p\pi}{2}}\left(A^4+e^{\frac{3p\pi}{2}}+A^5e^{\frac{3p\pi}{2}}+A^3p+A^2p^2+Ap^3+p^4\right)$$

$$=\frac{1}{5!}+\frac{A^5}{5!}+\frac{1}{5!}e^{-\frac{3p\pi}{2}}\frac{A^5-p^5}{A-p},$$

$$Y(6)=\frac{1}{720}e^{-\frac{3p\pi}{2}}\left(A^5+A^6e^{\frac{3p\pi}{2}}+A^4p+A^3p^2+A^2p^3+Ap^4+p^5\right)$$

$$=0+\frac{A^6}{6!}+\frac{1}{6!}e^{-\frac{3p\pi}{2}}\frac{A^6-p^6}{A-p},$$

$$\vdots$$

Then, we obtain the solution to (12):

$$y(t)=\left(\frac{t}{1!}-\frac{t^3}{3!}+\frac{t^5}{5!}+\cdots\right)+\sum_{k=0}^{\infty}\frac{A^k}{k!}t^k+e^{-\frac{3p\pi}{2}}\sum_{k=0}^{\infty}\frac{1}{k!}\frac{A^k-p^k}{A-p}t^k$$

$$=\sin(t)+e^{At}+\frac{e^{-\frac{3p\pi}{2}}}{A-p}\left(\sum_{k=0}^{\infty}\frac{A^k}{k!}t^k-\sum_{k=0}^{\infty}\frac{p^k}{k!}t^k\right)$$

$$=\sin(t)+e^{At}+\frac{e^{-\frac{3p\pi}{2}}}{A-p}\left(e^{At}-e^{pt}\right).$$

This is the analytic solution to (12). Particularly, if $A = p - e^{-\frac{3p\pi}{2}}$, then the above solution can be simplified to $y(t) = e^{pt} + \sin(t)$, which coincides with the definition on $\left[-\frac{3\pi}{2}, 0\right]$. So in this case, $y(t) = e^{pt} + \sin(t)$ is the analytic solution to (12) in the whole interval $[0, \infty]$.

As the last example, we apply the DTM based on the method of steps to solve a neutral delay differential equations.

Example 4.4 Consider the neutral delay differential equation

$$y'(x) = y'(x-1) - y(x), \quad 0 \le x \le 1 \tag{13}$$

with the initial function $y(x) = \sin(x)$, $x \le 0$

According to the idea of the method of steps, DDE (13) becomes

$$y'(x) = \cos(x-1) - y(x), \quad 0 \le x \le 1$$

Applying DTM to this equation, we have

$$(k+1)Y(k+1) = \frac{1}{k!}\cos\left(\frac{\pi k}{2} - 1\right) - Y(k)$$

From the initial function, we get $y(0) = 0$, so $Y(0) = 0$. With the help of Mathematica, we have

$$Y(1) = \cos(1), \quad Y(2) = \frac{1}{2!}(-\cos(1) + \sin(1)), \quad Y(3) = -\frac{\sin(1)}{3!}, \quad Y(4) = 0,$$
$$Y(5) = \frac{\cos(1)}{5!}, \quad Y(6) = \frac{1}{6!}(-\cos(1) + \sin(1)), \quad Y(7) = -\frac{\sin(1)}{7!}, \quad Y(8) = 0,$$
$$Y(9) = \frac{\cos(1)}{9!}, \quad Y(10) = \frac{-\cos(1) + \sin(1)}{10!}, \quad Y(11) = -\frac{\sin(1)}{11!}, \quad Y(12) = 0,$$
$$\cdots$$

Then, the solution to (13) is

$$
\begin{aligned}
y(x) &= \cos(1)\sum_{k=0}^{+\infty}\frac{x^{4k+1}}{(4k+1)!}+(\sin(1)-\cos(1))\sum_{k=0}^{+\infty}\frac{x^{4k+2}}{(4k+2)!}-\sin(1)\sum_{k=0}^{+\infty}\frac{x^{4k+3}}{(4k+3)!} \\
&= \cos(1)\frac{\sin(x)+\sinh(x)}{2}+(\sin(1)-\cos(1))\frac{-\cos(x)+\cosh(x)}{2}-\sin(1)\frac{-\sin(x)+\sinh(x)}{2} \\
&= \frac{1}{2}e^{-x}(\sin(1)-\cos(1))+\frac{1}{2}(\cos(1-x)-\sin(1-x)).
\end{aligned}
$$

5. CONCLUSION

Although the theory of differential transform method is not complete yet, it has been successfully applied to solve ordinary differential equations, partial differential equations, integral-differential equations, differential- algebraic equations and etc. In this paper, we apply DTM based on the method of steps to solve some delay differential equations, including neutral delay differential equations, successfully. Numerical experiments show that DTM is feasible and efficient for them. We believe that the operations of DTM presented in this paper also can be used to solve some partial delay differential equations (PDDEs), which is worth while studying in the future work.

ACKNOWLEDGEMENTS

Baoqing Liu, Xiaojian Zhou, Qikui Du This work is supported by the National Natural Science Foundation of China, contract/grant number 11371198 and 11401296, Jiangsu Provincial Natural Science Foundation of China, contact/grant no. BK20141008, Natural Science Fund for colleges and universities in Jiangsu Province contact/grant no. 14KJB110007, Jiangsu Provin- cial Key

Laboratory for Numerical Simulation of Large Scale Complex Systems contract/grant number 201305 and 201401.

REFERENCES

1. Pukhov, G.E. (1986) Differential Transformations and Mathematical Modelling of Physical Processes. Naukova Dumka, Kiev.

2. Zhou, J.K. (1986) Differential Transformation and Its Application for Electrical Circuits. Huazhong University Press, Wuhan. (In Chinese)

3. Ayaz, F. (2004) Solutions of the System of Differential Equations by Differential Transform Method. Applied Mathematics and Computation, 147, 547-567.

4. Kurnaz, A., Oturnaz, G. and Kiris, M.E. (2005) n-Dimensional Differential Transformation Method for Solving PDEs. International Journal of Computer Mathematics, 82, 369-380.

5. Ayaz, F. (2004) Applications of Differential Transform Method to Differential-Algebraic Equations. Applied Mathematics and Computation, 152, 649-657.

6. Liu, H. and Song, Y.Z. (2007) Differential Transform Method Applied to High Index Differential-Algebraic Equations. Applied Mathematics and Computation, 184, 748-753.

7. Abdel-Halim Hassan, I.H. (2002) On Solving Same Eigenvalue Problems by Using a Differential Transformation. Applied Mathematics and Computation, 127, 1-22.

8. Biazar, J., Eslami, M. and Islam, M.R. (2012) Differential Transform Method for Special Systems of Integral Equations. Journal of King Saud University (Science), 24, 211-214.

9. Shadia, M. (1992) Numerical Solution of Delay Differential and Neutral Differential Equations Using Spline Methods. Ph.D. Thesis, Assuit University, Assuit.

10. El-Hawary, H.M. and Mahmoud, S.M. (2003) Spline Collocation Methods for Solving Delay-Differential Equations. Applied Mathematics and Computation, 146, 359-372.

11. Evans, D.J. and Raslan, K.R. (2005) The Adomian Decomposition Method for Solving Delay Differential Equation. International Journal of Computer Mathematics, 82, 49-54.

12. Vanani, S.K. and Aminataei, A. (2008) On the Numerical Solution of Neutral Delay Differential Equations Using Multiquadric Approximation Scheme. Bulletin of the Korean Mathematical Society, 45, 663-670.

13. Karako, F. and Bereketoğlu, H. (2009) Solutions of Delay Differential Equations by Using Differential Transform Method. International Journal of Computer Mathematics, 86, 914-923.

14. Lainscsek, C. and Sejnowski, T.J. (2015) Delay Differential Analysis of Time Series. Neural Computation, 27, 594- 614.

15. Abazari, R. and Kilicman, A. (2014) Application of Differential Transform Method on Nonlinear Integro-Differential Equations with Proportional Delay. Neural Computing and Applications, 24, 391-397.

16. Driver, R.D. (1977) Ordinary and Delay Differential Equations. Applied Mathematical Sciences, 20.

17. Odibat, Z.M., et al. (2010) A Multi-Step Differential Transform Method and Application to Non-Chaotic or Chaotic Systems. Computers & Mathematics with Applications, 59, 1462-1472. http://dx.doi.org/10.1016/j.camwa.2009.11.005

18. Hairer, E., et al. (1993) Solving Ordinary Differential Equations I: Nonstiff Problems. 2nd Edition, Springer-Verlag, Berlin.

CHAPTER 7

New Oscillation Results for Forced Second Order Differential Equations with Mixed Nonlinearities

Ercan Tunç, Adil Kaymaz

Department of Mathematics, Faculty of Arts and Sciences, Gaziosmanpaşa University, Tokat, Turkey

ABSTRACT

Some new oscillation criteria are given for forced second order differential equations with mixed nonlinearities by using the generalized variational principle and Riccati technique. Our results generalize and extend some known oscillation results in the literature.

KEYWORDS

Generalized variational principle; variational principle; second order differential equations; Riccati transformation; oscillation

1. INTRODUCTION

The oscillatory behavior of second order differential equations has a major role in the theory of differential equations. It has been shown that many real world problems can be modelled, in particular, by half linear differential equations which can be regarded as a natural generalization of linear differential equations [1- 14]. A considerable amount of research has also been done on quasi-linear [15-18] and nonlinear second order differential equations [19-23].

In this paper, we investigate the oscillatory behavior of second order forced differential equation with mixed nonlinearities.

$$\left(r(t)|x'(t)|^{\alpha-1}x'(t)\right)' + p(t)|x(t)|^{\alpha-1}x(t)$$

$$+\sum_{j=1}^{m} q_j(t)|x(t)|^{\beta_j-1}x(t) = e(t),\ t \ge t_0, \tag{1}$$

where $r \in C([t_0,\infty),(0,\infty))$, $p, q_j\ (1 \le j \le m), e \in C([t_0,\infty),(-\infty,\infty))$ and $0 < \alpha < \beta_1 < \beta_2 < \cdots < \beta_m$ are real numbers, $p, q_j\ \ (1 \le j \le m)$ and e might alternate signs.

By a solution of Equation (1), we mean a function $x(t) \in C^1([T_x,\infty),(-\infty,\infty))$, where $T_x \ge t_0$ depends on the particular solution, which has the property that $r(t)|x'(t)|^{\alpha-1}x'(t) \in C^1[T_x,\infty)$ and satisfies Equation (1). We restrict our attention to the nontrivial solutions $x(t)$ of Equation (1) only, i.e., to solutions $x(t)$ such that $\sup\{|x(t)| : t \ge T\} > 0$ for all $T \ge T_x$. A nontrivial solution of (1) is oscillatory if it has arbitrarily large zeros, otherwise, it is called non-oscillatory. Equation (1) is said to be oscillatory if all its nontrivial solutions are oscillatory.

Equation (1) and its special cases such as the linear differential equation

$$\left(r(t)x'(t)\right)' + q(t)x(t) = e(t), \tag{2}$$

the half-linear differential equation

$$\left(r(t)\left|x'(t)\right|^{\alpha-1} x'(t)\right)' + q(t)\left|x(t)\right|^{\alpha-1} x(t) = e(t) \tag{3}$$

and the quasi-linear differential equation

$$\left(r(t)\left|x'(t)\right|^{\alpha-1} x'(t)\right)' + q(t)\left|x(t)\right|^{\beta-1} x(t) = e(t) \tag{4}$$

have been extensively studied by numerous authors with different methods (see, for example, [1-5,15-19] and the references quoted therein).

In 1999, Wong [1] proved the following theorem by making use of the "oscillatory intervals" of e(t) and Leighton's variational principle (see [10]) for (2).

Theorem 1.1. Suppose that for any $T \geq t_0$, there exist $T \leq s_1 < t_1 \leq s_2 < t_2$ such that

$$e(t)\begin{cases} \leq 0, & t \in [s_1, t_1] \\ \geq 0, & t \in [s_2, t_2]. \end{cases} \tag{5}$$

Denote

$$D(s_i, t_i) = \left\{u \in C^1[s_i, t_i] : u(t) \not\equiv 0, u(s_i) = u(t_i) = 0\right\},$$
$i = 1, 2.$

If there exist $u \in D(s_i, t_i)$ such that

$$Q_i(u) = \int_{s_i}^{t_i} \left[q(t)u^2(t) - r(t)\left(u'(t)\right)^2 \right] \mathrm{d}t \geq 0, \ i = 1, 2 \tag{6}$$

then Equation (2) is oscillatory.

Afterwards, in 2002, the authors of [2] extended Wong's results, using a similar method, to Equation (3) as follows.

Theorem 1.2. Suppose that for any $T \geq t_0$, there exist $T \leq s_1 < t_1 \leq s_2 < t_2$ such that (5) holds. Let

$D(s_i, t_i) = \{u \in C^1[s_i, t_i] : u(t) \not\equiv 0, u(s_i) = u(t_i) = 0\}$,
$i = 1, 2$.

If there exist $H \in D(s_i, t_i)$ and a positive, nondecreasing function $\phi \in C^1([t_0, \infty))$ such that

$$\int_{s_i}^{t_i} H^2(t)\phi(t)q(t)\mathrm{d}t$$
$$> K \int_{s_i}^{t_i} \frac{r(t)\phi(t)}{|H(t)|^{\alpha-1}} \left(2|H'(t)| + |H(t)| \frac{\phi'(t)}{\phi(t)} \right)^{\alpha+1} \mathrm{d}t, \tag{7}$$

for i = 1, 2, where $K = (1/(\alpha+1))^{\alpha+1}$, then (3) is oscillatory.

Later, in 2007, Zheng and Meng [16], considering a more general equation (4), improved the paper [2] and showed that the results obtained in [2] for Equation (3) can not be applied to the case $\alpha > 1$. The main result of Zheng and Meng [16] is the following.

Theorem 1.3. Assume that for any $T \geq t_0$, there exist $T \leq s_1 < t_1 \leq s_2 < t_2$ such that (5) holds. Let

$$D(s_i,t_i) = \{u \in C^1[s_i,t_i] : u^{\alpha+1}(t) > 0,\ t \in (s_i,t_i) \text{ and } u(s_i) = u(t_i) = 0\} \text{ for } i = 1,2.$$

Suppose that there exist $H \in D(s_i,t_i)$ and a positive, nondecreasing function $\phi \in C^1([t_0,\infty),(-\infty,\infty))$ such that

$$Q_i^{\phi}(H) := \int_{s_i}^{t_i} \phi(t)\left[Q(t)H^{\alpha+1}(t) - r(t)\left(|H'(t)| + \frac{H(t)\phi'(t)}{(\alpha+1)\phi(t)}\right)^{\alpha+1}\right]\mathrm{d}t > 0 \tag{8}$$

for i =1, 2. Then Equation (4) is oscillatory, where

$$Q(t) = \alpha^{-\alpha/\beta}\beta((\beta-\alpha))^{(\alpha-\beta)/\beta}\left[q(t)\right]^{\alpha/\beta}|e(t)|^{(\beta-\alpha)/\beta},\ 0 < \alpha \leq \beta \tag{9}$$

with the convention that $0^0 = 1$.

Also, in [2009], Zheng et al. [17] extended the results obtained for Equation (4) to Equation (1) as follows.

Theorem 1.4. Assume that for any $T \geq t_0$, there exist $T \leq s_1 < t_1 \leq s_2 < t_2$ such that $q_j(t) \geq 0 (1 \leq j \leq m)$ for $t \in [s_1,t_1] \cup [s_2,t_2]$ and (5) holds. Let

$$D(s_i,t_i) = \{u \in C^1[s_i,t_i] : u^{\alpha+1}(t) > 0, t \in (s_i,t_i) \text{ and } u(s_i) = u(t_i) = 0\} \text{ for } i = 1,2$$

If there exist $H \in D(s_i,t_i)$ and a positive function $\phi \in C^1([t_0,\infty),R)$ such that

$$\int_{s_i}^{t_i} \phi(t)\left[\left(p(t) + \sum_{j=1}^{m} Q_j(t)\right)H^{\alpha+1}(t) - r(t)\left(|H'(t)| + \frac{H(t)\phi'(t)}{(\alpha+1)\phi(t)}\right)^{\alpha+1}\right]\mathrm{d}t > 0 \tag{10}$$

for i = 1, 2. Then Equation (1) is oscillatory, where

$$Q_j(t)=\alpha^{-\alpha/\beta_j}\beta_j\left[m(\beta_j-\alpha)\right]^{(\alpha-\beta_j)/\beta_j}\left[q_j(t)\right]^{\alpha/\beta_j}|e(t)|^{(\beta_j-\alpha)/\beta_j},\quad 1\le j\le m \tag{11}$$

with the convention that $0^0=1$.

Recently, Shao [15] generalized the results of Zheng and Meng [16] by using the generalized variational principle due to Komkov [24] and gave the following result for Equation (4).

Theorem 1.5. Assume that, for any $T\ge t_0$, there exist $T\le s_1<t_1\le s_2<t_2$ such that (1.5) holds. Let $u\in C^1[s_i,t_i]$, and nonnegative functions G_1,G_2 satisfying $G_i(u(s_i))=G_i(u(t_i))=0$, $g_i(u)=G_i'(u)$ are continuous and $(g_i(u(t)))^{\alpha+1}\le(\alpha+1)^{\alpha+1}G_i^{\alpha}(u(t))$

for $t\in[s_i,t_i]$, i = 1, 2. If there exists a positive function $\phi\in C^1([t_0,\infty),R)$ such that

$$Q_i^{\phi}(u)$$
$$:=\int_{s_1}^{t_1}\phi\left[QG_i(u)-r\left(|u'|+\frac{G_i^{1/(\alpha+1)}(u)|\phi'|}{(\alpha+1)\phi}\right)^{\alpha+1}\right](t)\,\mathrm{d}t>0 \tag{12}$$

for i = 1, 2, then Equation (4) is oscillatory, where $Q(t)$ is the same as (9).

Motivated by the above theorems we propose some new oscillation results by employing the generalized variational principle and Riccati technique for Equation (1). Our results extend and generalize some known results in the literature. We now state our main results and several remarks.

2. NEW OSCILLATION RESULTS

In order to prove our results we use the following wellknown inequality which is presented by Hardy et al. [25].

Lemma 2.1. (see [25]). If X and Y are nonnegative, then

$$X^{\gamma}+(\gamma-1)Y^{\gamma}\geq\gamma XY^{\gamma-1},\ \gamma>1, \tag{13}$$

where equality holds if and only if $X=Y.$

Theorem 2.1. Assume that, for any $T\geq t_0$, there exist $T\leq s_1<t_1\leq s_2<t_2$ such that

$q_j(t)\geq 0(1\leq j\leq m)$ for $t\in[s_1,t_1]\cup[s_2,t_2]$ and (5)

holds. Let $u\in C^1[s_i,t_i]$ and nonnegative functions

G_1, G_2 satisfying

$$G_i(u(s_i))=G_i(u(t_i))=0,\ g_i(u)=G_i'(u)$$

are continuous and

$$(g_i(u(t)))^{\alpha+1}\leq(\alpha+1)^{\alpha+1}G_i^{\alpha}(u(t))$$

for $t\in[s_i,t_i]$, i = 1, 2. If there exists a positive function $\phi\in C^1([t_0,\infty),R)$ such that

$$\int_{s_i}^{t_i}\phi(t)\left[\left(p(t)+\sum_{j=1}^{m}Q_j(t)\right)G_i(u(t))-r(t)\left(|u'(t)|+\frac{G_i^{1/(\alpha+1)}(u(t))|\phi'(t)|}{(\alpha+1)\phi(t)}\right)^{\alpha+1}\right]dt>0 \tag{14}$$

for i =1, 2, then Equation (1) is oscillatory, where $Q_j(t)$ is the same as (11).

Proof. Suppose that $x = x(t)$ is a nonoscillatory solution of Equation (1). Then, there exists a $T_0 \geq t_0$ such that $x(t) \neq 0$ for all $t \geq T_0$. Without loss of generality, we may assume that $x(t) > 0$ for all $t \geq T_0$. We introduce the Ricccati transformation

$$w(t) = \phi(t) \frac{r(t)|x'(t)|^{\alpha-1} x'(t)}{|x(t)|^{\alpha-1} x(t)}, \quad t \geq T_0. \tag{15}$$

Differentiating (15) and using (1), we obtain, for all $t \geq T_0$,

$$\begin{aligned} w'(t) &= \frac{\phi'(t)}{\phi(t)} w(t) \\ &\quad + \phi(t) \left[\frac{\left(-p(t)|x(t)|^{\alpha-1} x(t) - \sum_{j=1}^{m} q_j(t)|x(t)|^{\beta_j-1} x(t) + e(t) \right) |x(t)|^{\alpha-1} x(t) - r(t)|x'(t)|^{\alpha-1} x'(t) \left(|x(t)|^{\alpha-1} x(t) \right)'}{\left(|x(t)|^{\alpha-1} x(t) \right)^2} \right] \\ &= \frac{\phi'(t)}{\phi(t)} w(t) - \phi(t) p(t) - \phi(t) \left[\sum_{j=1}^{m} q_j(t)|x(t)|^{\beta_j-\alpha} - \frac{e(t)}{|x(t)|^{\alpha-1} x(t)} \right] - \alpha \frac{|w(t)|^{(\alpha+1)/\alpha}}{(\phi(t) r(t))^{1/\alpha}}. \end{aligned} \tag{16}$$

By the assumption, we can choose $s_1, t_1 \geq T_0$ so that $e(t) \leq 0$ on the interval $I_1 = [s_1, t_1]$ with $s_1 < t_1$. As in [18], for given $t \in I_1$, set

$$F_j(s) = q_j(t) s^{\beta_j - \alpha} - \frac{e(t)}{m s^{\alpha}}, \quad 1 \leq j \leq m, \quad s > 0.$$

It is easy to verify that

$$F_j'\left(s_j^*\right)=0 \text{ and } F_j''\left(s_j^*\right)>0,$$

$$\text{where } s_j^*=\left[\frac{-\alpha e(t)}{m\left(\beta_j-\alpha\right)q_j(t)}\right]^{1/\beta_j}.$$

So $F_j(s)$ obtains its minimum on s_j^* and

$$F_j(s)\geq F_j\left(s_j^*\right)=\alpha^{-\alpha/\beta_j}\beta_j\left[m\left(\beta_j-\alpha\right)\right]^{\left(\alpha-\beta_j\right)/\beta_j}\left[q_j(t)\right]^{\alpha/\beta_j}\left|e(t)\right|^{\left(\beta_j-\alpha\right)/\beta_j}=Q_j(t). \tag{17}$$

Then, by using (17) in (16), we get

$$\phi(t)\left(p(t)+\sum_{j=1}^{m}Q_j(t)\right)\leq -w'(t)+\frac{\phi'(t)}{\phi(t)}w(t)-\alpha\frac{\left|w(t)\right|^{(\alpha+1)/\alpha}}{\left(\phi(t)r(t)\right)^{1/\alpha}}. \tag{18}$$

Multiplying $G_1(u(t))$ through (18) and integrating over I_1, we have

$$\int_{s_1}^{t_1}\phi(t)\left(p(t)+\sum_{j=1}^{m}Q_j(t)\right)G_1(u(t))dt\leq-\int_{s_1}^{t_1}G_1(u(t))w'(t)dt+\int_{s_1}^{t_1}G_1(u(t))\frac{\phi'(t)}{\phi(t)}w(t)dt-\alpha\int_{s_1}^{t_1}G_1(u(t))\frac{\left|w(t)\right|^{(\alpha+1)/\alpha}}{\left(\phi(t)r(t)\right)^{1/\alpha}}dt. \tag{19}$$

By integration by parts and using the fact that $G_1(u(s_1))=G_1(u(t_1))=0,$ we have

$$\int_{s_1}^{t_1}G_1(u(t))w'(t)dt=G_1(u(t))w(t)\Big|_{s_1}^{t_1}-\int_{s_1}^{t_1}G_1'(u(t))u'(t)w(t)dt=-\int_{s_1}^{t_1}g_1(u(t))u'(t)w(t)dt. \tag{20}$$

In view of (19) and (20), we conclude that

$$\int_{s_1}^{t_1}\phi(t)\left(p(t)+\sum_{j=1}^{m}Q_j(t)\right)G_1(u(t))\mathrm{d}t$$
$$\leq\int_{s_1}^{t_1}\left[g_1(u(t))u'(t)+G_1(u(t))\frac{\phi'(t)}{\phi(t)}\right]w(t)\mathrm{d}t-\alpha\int_{s_1}^{t_1}G_1(u(t))\frac{|w(t)|^{(\alpha+1)/\alpha}}{(\phi(t)r(t))^{1/\alpha}}\mathrm{d}t$$
$$\leq\int_{s_1}^{t_1}\left[|g_1(u(t))||u'(t)|+G_1(u(t))\frac{|\phi'(t)|}{\phi(t)}\right]|w(t)|dt-\alpha\int_{s_1}^{t_1}G_1(u(t))\frac{|w(t)|^{(\alpha+1)/\alpha}}{(\phi(t)r(t))^{1/\alpha}}\mathrm{d}t$$
$$\leq(\alpha+1)\int_{s_1}^{t_1}\left[G_1^{\alpha/(\alpha+1)}(u(t))|u'(t)|+G_1(u(t))\frac{|\phi'(t)|}{(\alpha+1)\phi(t)}\right]|w(t)|\mathrm{d}t-\alpha\int_{s_1}^{t_1}G_1(u(t))\frac{|w(t)|^{(\alpha+1)/\alpha}}{(\phi(t)r(t))^{1/\alpha}}\mathrm{d}t. \tag{21}$$

Let

$$X=\left[\frac{\alpha}{(\phi(t)r(t))^{1/\alpha}}\right]^{\alpha/(\alpha+1)}G_1^{\alpha/(\alpha+1)}(u(t))|w(t)|,\ \gamma=\frac{\alpha+1}{\alpha},\ Y=(\alpha\phi(t)r(t))^{\alpha/(\alpha+1)}\left[|u'(t)|+\frac{G_1^{1/(\alpha+1)}(u(t))|\phi'(t)|}{(\alpha+1)\phi(t)}\right]^{\alpha}.$$

According to Lemma 2.1, we obtain for $t\in[s_1,t_1]$

$$(\alpha+1)\left[G_1^{\alpha/(\alpha+1)}(u(t))|u'(t)|+G_1(u(t))\frac{|\phi'(t)|}{(\alpha+1)\phi(t)}\right]|w(t)|-\alpha G_1(u(t))\frac{|w(t)|^{(\alpha+1)/\alpha}}{(\phi(t)r(t))^{1/\alpha}}$$
$$\leq\frac{1}{\alpha}(\alpha\phi(t)r(t))\left[|u'(t)|+\frac{G_1^{1/(\alpha+1)}(u(t))|\phi'(t)|}{(\alpha+1)\phi(t)}\right]^{\alpha+1}=\phi(t)r(t)\left[|u'(t)|+\frac{G_1^{1/(\alpha+1)}(u(t))|\phi'(t)|}{(\alpha+1)\phi(t)}\right]^{\alpha+1}.$$

Therefore, (21) yields

$$\int_{s_1}^{t_1}\phi(t)\left(p(t)+\sum_{j=1}^{m}Q_j(t)\right)G_1(u(t))\mathrm{d}t\leq\int_{s_1}^{t_1}\phi(t)r(t)\left[|u'(t)|+\frac{G_1^{1/(\alpha+1)}(u(t))|\phi'(t)|}{(\alpha+1)\phi(t)}\right]^{\alpha+1}\mathrm{d}t,$$

which contradicts the assumption (14) for $i=1$.

When $x(t)$ is a negative solution for $t\geq T_0>t_0$, we may employ the fact that $e(t)\geq 0$ on $I_2=[s_2,t_2]$ to reach a similar contradiction. Therefore, any solution $x(t)$ can be neither eventually positive nor eventually negative. Hence, any solution is oscillatory. This completes the proof of Theorem 2.1.

If $p(t)\equiv 0$ and $m=1$, then Equation (1) reduces to Equation (4). Thus by Theorem 2.1, we have the following oscillation result:

Corollary 2.1. Assume that, for any $T\geq t_0$, there exist $T\leq s_1<t_1\leq s_2<t_2$ such that (5) holds. Let $u\in C^1[s_i,t_i]$, and nonnegative functions G_1,G_2 satisfying $G_i(u(s_i))=G_i(u(t_i))=0$, $g_i(u)=G_i'(u)$ are continuous and $(g_i(u(t)))^{\alpha+1}\leq(\alpha+1)^{\alpha+1}G_i^{\alpha}(u(t))$ for $t\in[s_i,t_i]$ for i =1, 2. If there exists a positive function $\phi\in C^1([t_0,\infty),R)$ such that

$$\int_{s_i}^{t_i}\phi(t)\left[Q(t)G_i(u(t))-r(t)\left(|u'(t)|+\frac{G_i^{1/(\alpha+1)}(u(t))|\phi'(t)|}{(\alpha+1)\phi(t)}\right)^{\alpha+1}\right]dt>0 \tag{22}$$

for i = 1, 2, then Equation (4) is oscillatory, where $Q(t)$ is the same as (9).

Remark 1. Corollary 2.1 shows that Theorem 2.1 is a generalization of Theorem 1.5.

Remark 2. Let $G_1(u)=G_2(u)=u^{\alpha+1}$ in Corollary 2.1, then our main Theorem 2.1 reduces to Theorem 1.3.

Remark 3. If we choose $G_1(u)=G_2(u)=u^{\alpha+1}$ in Theorem 2.1, then we obtain Theorem 1.4.

Remark 4. If we choose $G_1(u)=G_2(u)=u^{\alpha+1}$ and $\phi(t)\equiv 1$ in Theorem 2.1, then we obtain Corollary 2.3 of Paper [17].

Remark 5. If we choose $G_1(u)=G_2(u)=u^{\alpha+1}$ and $\phi(t)\equiv 1$ in Corollary 2.1, then we obtain Corollary 2.3 of paper [16].

Remark 6. Let

$$G_1(u) = G_2(u) = u^{\alpha+1},\ q_j(t) \equiv 0, 1 \le j \le m$$

and $\phi(t) \equiv 1$ in Theorem 2.1, then Theorem 2.1 is a generalization of Theorem 1.1.

Remark 7. Let $q_j(t) \equiv 0, (1 \le j \le m)$. If we choose $G_1(u) = G_2(u) = u^{\alpha+1}$ in Theorem 2.1, then Theorem 2.1 improves Theorem 1.2, since the positive constant α in Theorem 2.1 can be chosen as any number lying in $(0,\infty)$.

Remark 8. If the condition (5) in Theorem 2.1 and Corollary 2.1 is replaced by

$$e(t)\begin{cases} \ge 0, t \in [s_1, t_1] \\ \le 0, t \in [s_2, t_2] \end{cases}$$

then the results given in this paper are still valid.

3. EXAMPLES

Example 3.1. Consider

$$\left(\gamma t^{\lambda/3} |x'(t)|^{\alpha-1} x'(t)\right)' + t^{\lambda/3} |x(t)|^{\alpha-1} x(t) + t^{\lambda} |x(t)|^{\beta-1} x(t) = -\sin^3 t, \tag{23}$$

for $t \ge 1$, where $\gamma, \lambda > 0$ are constants. Let $\alpha = 1$

and $\beta = 3$, so $Q(t) = \left(3/\sqrt[3]{4}\right)t^{\lambda/3}\sin^2 t$. The zeros of forcing term $-\sin^3 t$ are $n\pi$. For any $T \geq 1$, we choose n sufficiently large so that $n\pi = 2k\pi \geq T$, $s_1 = 2k\pi$ and $t_1 = (2k+1)\pi$. Letting $u(t) = \sin t \geq 0$, $G_1(u) = u^2\exp(-u)$ (it is easy to verify that $\left(G_1'(u)\right)^2 \leq 4G_1(u)$ for $u \geq 0$), $\phi(t) = t^{-\lambda/3}$ then we obtain

$$\int_{s_1}^{t_1}\phi(t)\left(p(t)+Q(t)\right)G_i\left(u(t)\right)\mathrm{d}t = \int_{2k\pi}^{(2k+1)\pi}\left(\sin^2 t + \left(3/\sqrt[3]{4}\right)\sin^4 t\right)\exp(-\sin t)\mathrm{d}t \geq \frac{\pi}{2e}\left(1+\frac{9}{4\sqrt[3]{4}}\right),$$

and

$$\int_{s_1}^{t_1}\phi(t)r(t)\left(\left|u'(t)\right| + \frac{G_i^{1/(\alpha+1)}\left(u(t)\right)\left|\phi'(t)\right|}{(\alpha+1)\phi(t)}\right)^{\alpha+1}\mathrm{d}t = \int_{2k\pi}^{(2k+1)\pi}\gamma\left(\left|\cos t\right| + \frac{\lambda\sin t\exp(-\sin t/2)}{6t}\right)^2\mathrm{d}t < \gamma\left(1+\frac{\lambda}{6}\right)^2\pi.$$

Therefore, Equation (14) is satisfied for i = 1 provided that $0 < \gamma < \frac{1}{2e}\left(1+\frac{9}{4\sqrt[3]{4}}\right)\Big/\left(1+\frac{\lambda}{6}\right)^2$. In a similar way, for $s_2 = (2k+1)\pi$ and $t_2 = (2k+2)\pi$, we choose $u(t) = \sin t \leq 0$, $G_2(u) = u^2\exp(u)$ (it is easy to verify that $\left(G_2'(u)\right)^2 \leq 4G_2(u)$ for $u \leq 0$) so that that (14) is valid for i = 2. Thus (23) is oscillatory for

$$0 < \gamma < \frac{1}{2e}\left(1+\frac{9}{4\sqrt[3]{4}}\right)\Big/\left(1+\frac{\lambda}{6}\right)^2$$

by Theorem 2.1.

Example 3.2. Consider the following forced quasilinear differential equation

$$\left[\gamma(2+\cos t)t^{-\lambda/5}\left|x'\right|^{\alpha-1}x'\right]' + t^{-\lambda/5}\exp(\sin t/5)\left|x\right|^{\alpha-1}x + t^{-\lambda}\exp(\sin t)\left|x\right|^{\beta-1}x = -\sin^5 t \tag{24}$$

for $t \geq 1$, where $\gamma, \lambda > 0$ are constants. Let $\alpha = 1$

and $\beta = 5$, so $Q(t) = \frac{5}{\sqrt[5]{4^4}} t^{-\lambda/5} \exp(\sin t/5) \sin^4 t.$ The zeros of forcing term $-\sin^5 t$ are $n\pi$. For any $T \geq 1$,

we choose n sufficiently large so that $n\pi = 2k\pi \geq T$, $s_1 = 2k\pi$ and $t_1 = (2k+1)\pi.$ Letting

$G_1(u) = u^2 \exp(-u)$ $u(t) = \sin t \geq 0$, $\phi(t) = t^{\lambda/5}$ then we obtain

$$\int_{s_1}^{t_1} \phi(t)\big(p(t)+Q(t)\big)G_i\big(u(t)\big)\mathrm{d}t = \int_{2k\pi}^{(2k+1)\pi} \left(\sin^2 t + \frac{5}{\sqrt[5]{4^4}}\sin^6 t\right)\exp\left(\frac{-4\sin t}{5}\right)\mathrm{d}t$$
$$\geq \frac{1}{e^{4/5}}\int_0^{\pi}\left(\sin^2 t + \frac{5}{\sqrt[5]{4^4}}\sin^6 t\right) = \frac{\pi}{16e^{4/5}}\left(8+\frac{25}{\sqrt[5]{4^4}}\right)$$

and

$$\int_{s_1}^{t_1} \phi(t)r(t)\left(|u'(t)| + \frac{G_i^{1/(\alpha+1)}\big(u(t)\big)|\phi'(t)|}{(\alpha+1)\phi(t)}\right)^{\alpha+1}\mathrm{d}t = \int_{2k\pi}^{(2k+1)\pi}\gamma(2+\cos t)\left(|\cos t| + \frac{\lambda\sin t\exp(-\sin t/2)}{10t}\right)^2\mathrm{d}t$$
$$< \int_0^{\pi} 3\gamma(1+\lambda/10)^2\mathrm{d}t = 3\gamma(1+\lambda/10)^2\,\pi.$$

Therefore, Equation (14) is satisfied for i = 1 provided that $0 < \gamma < \frac{\delta}{3(1+\lambda/10)^2}$, where $\delta = \frac{1}{16e^{4/5}}\left(8+\frac{25}{\sqrt[5]{4^4}}\right).$

In a similar way, for $s_2 = (2k+1)\pi$ and $t_2 = (2k+2)\pi$, we choose $u(t) = \sin t \leq 0$, $G_2(u) = u^2\exp(u)$ so that (14) is valid for i = 2. Thus (24) is oscillatory for $0 < \gamma < \delta/3(1+\lambda/10)^2$ by Theorem 2.1.

4. CONCLUSION

The oscillatory behavior of many different kinds of differential equations has been investigated and a great deal of results has been obtained in the literature. In this article, we generalized the results obtained in [16,17] and extended the results of Shao [15] by using the generalized variational principle and Riccati tecnique. In a similar way, the results obtained for Equation (1) can be extended to a more general class of differential equations.

ACKNOWLEDGEMENTS

The authors would like to express sincere thanks to the anonymous referee for her/his invauable corrections, comments and suggestions on the paper.

REFERENCES

1. J. S. W. Wong, "Oscillation Criteria for a Forced SecondOrder Linear Differential Equation," Journal of Mathematical Analysis and Applications, Vol. 231, No. 1, 1999, pp. 235-240.

2. W. T. Li and S. S. Cheng, "An Oscillation Criterion for Nonhomogeneous Half-Linear Differential Equations," Applied Mathematics Letters, Vol. 15, No. 3, 2002, pp. 259-263.

3. J. V. Manojlovic, "Oscillation Criteria for Second-Order Half-Linear Differential Equations," Mathematical and Computer Modelling, Vol. 30, No. 5-6, 1999, pp. 109- 119.

4. Q. R. Wang, "Oscillation and Asymptotics for SecondOrder Half-Linear Differential Equations," Applied Mathematics and Computation, Vol. 122, No. 2, 2001, pp. 253- 266.

5. Q. R. Wang and Q. G. Yang "Interval Criteria for Oscillation of Second-Order Half-Linear Differential Equations," Journal of Mathematical Analysis and Applications, Vol. 291, No. 1, 2004, pp. 224-236.

6. J. Jaros and T. Kusano, "A Picone Type Identity for Second Order Half-Linear Differential Equations," Acta Mathematica Universitatis Comenianae, Vol. 68, No. 1, 1999, pp. 137-151.

7. A. Elbert, "A Half-Linear Second Order Differential Equation," Colloquia Mathematica Societatis Janos Bolyai: Qualitative Theory of Differential Equations, Szeged, 1979, pp. 153-180.

8. A. Wintner, "A Criterion of Oscillatory Stability," Quarterly of Applied Mathematics, Vol. 7, 1949, pp. 115-117.

9. I. V. Kamenev, "An Integral Criterion for Oscillation of Linear Differential Equations of Second Order," Matematicheskie Zametki Vol. 23, No. 2, 1978, pp. 249-251.

10. W. Leighton, "Comparison Theorems for Linear Differential Equations of Second Order," Proceedings of the American Mathematical Society, Vol. 13, 1962, pp. 603- 610.

11. Q. Kong, "Interval Criteria for Oscillation of SecondOrder Linear Ordinary Differential Equation," Journal of Mathematical Analysis and Applications, Vol. 229, No. 1, 1999, pp. 258-270.

12. H. J. Li and C. C. Yeh, "Sturm Comparison Theorem for Half-Linear Second Order Differential Equations," Proceedings of

the Royal Society of Edinburgh, Vol. A125, 1995, pp. 1193-1240.

13. O. Došlý and P. Řehák, "Half-Linear Differential Equations," North-Holland Mathematics Studies, Vol. 202, Elsevier Science, Amsterdam, 2005.

14. R. P. Agarwal, S. R. Grace and D. O'Regan, "Oscillation Theory for Second Order Linear, Half-Linear, Superlinear Dynamic Equations," Kluver, Dordrecht, 2002.

15. J. Shao, "A New Oscillation Criterion for Forced SecondOrder Quasi-Linear Differential Equations," Discrete Dynamics in Nature and Society, Vol. 2011, Hindawi Publishing Corporation, New York, pp. 1-8.

16. Z. Zheng and F. Meng, "Oscillation Criteria for Forced Second Order Quasi-Linear Differential Equations," Mathematical and Computer Modelling, Vol. 45, No. 1-2, 2007, pp. 215-220.

17. Z. Zheng, X. Wang and H. Han, "Oscillation Criteria for Forced Second Order Differential Equations with Mixed Nonlinearities," Applied Mathematics Letters, Vol. 22, No. 7, 2009, pp. 1096-1101.

18. J. Jaros, T. Kusano and N. Yoshida, "Generalized Picone's Formula and Forced Oscillation in Quasi-Linear Differential Equations of the Second Order," Archivum Mathematicum, Vol. 38, No. 1, 2002, pp. 53-59.

19. J. Shao and F. Meng, "Generalized Variational Principles on Oscillation for Nonlinear Nonhomogeneous Differential Equations," Abstract and Applied Analysis, Vol. 2011, 2011, pp. 1-10.

20. Q. Yang, "Interval Oscillation Criteria for a Forced Second Order Nonlinear Ordinary Differential Equations with

Oscillatory Potential," Applied Mathematics and Computation, Vol. 135, No. 1, 2003, pp. 49-64.

21. D. Çakmak and A. Tiryaki, "Oscillation Criteria for Certain Forced Second-Order Nonlinear Differential Equations," Applied Mathematics Letters, Vol. 17, No. 3, 2004, pp. 275-279.

22. E. Tunç, "A Note on the Oscillation of Second-Order Differential Equations with Damping," Journal of Computational Analysis and Applications, Vol. 12, No. 2, 2010, pp. 444-453.

23. E. Tunç, "Interval Oscillation Criteria for Certain Forced Second-Order Differential Equations," Carpathian Journal of Mathematics, Vol. 28, No. 1, 2012, in Press.

24. V. Komkov, "A Generalization of Leighton's Variational Theorem," Applicable Analysis: An International Journal, Vol. 2, No. 4, 1972, pp. 377-383.

25. G. H. Hardy, J. E. Littlewood and G. Polya, "Inequalities," 2nd Edition, Cambridge University Press, Cambridge, 1988.

CHAPTER 8

An Integral Collocation Approach Based on Legendre Polynomials for Solving Riccati, Logistic and Delay Differential Equations

M. M. Khader[1,2], A. M. S. Mahdy[3], M. M. Shehata[3]

[1] Department of Mathematics and Statistics, College of Science, Al-Imam Mohammad Ibn Saud Islamic University (IMSIU), Riyadh, KSA

[2] Department of Mathematics, Faculty of Science, Benha University, Benha, Egypt

[3] Department of Mathematics, Faculty of Science, Zagazig University, Zagazig, Egypt

ABSTRACT

In this paper, we propose and analyze some schemes of the integral collocation formulation based on Legendre polynomials. We implement these formulae to solve numerically Riccati, Logistic and delay differential equations with variable coefficients. The properties of the Legendre polynomials are used to reduce the proposed problems to the solution of non-linear system of algebraic equations using Newton iteration method. We give numerical results to satisfy the accuracy and the applicability of the proposed schemes.

KEYWORDS

Integral Collocation Formulation, Spectral Method, Riccati, Logistic and Delay Differential Equations

1. INTRODUCTION

It is well known that the ordinary differential equations (ODEs) have been the focus of many studies due to their frequent appearance in various applications, such as in fluid mechanics, viscoelasticity, biology, physics and engineering applications, for more details, for example [1] -[5] . Consequently, considerable attention has been given to the efficient numerical solutions of ODEs of physical interest, because it is difficult to find exact solutions. Different numerical methods have been proposed in the literature for solving ODEs [6] -[13] .

The Riccati differential equation (RDE) is named after the Italian Nobleman Count Jacopo Francesco Riccati (1676-1754). The book of Reid [14] contains the fundamental theories of Riccati equation, with applications to random processes, optimal control, and diffusion problems. Besides important engineering science applications that today are considered classical, such as stochastic realization theory, optimal control, robust stabilization, and network synthesis, the newer applications include such areas as financial mathematics [15]. The solution of this equation can be reached using classical numerical methods such as, the forward Euler method and RungeKutta method. Bahnasawi et al. [16] presented the usage of Adomian decomposition method to solve the nonlinear RDE in an analytic form. Tan and Abbasbandy [17] employed the analytic technique called homotopy analysis method to solve the quadratic RDE.

The Logistic model can be obtained by applying the derivative operator on the Logistic equation. The model is initially published by Pierre Verhulst in 1838 [18] . The continuous Logistic model is described by first order ODE. The discrete Logistic model is simple iterative equation that reveals the chaotic property in certain regions [19]. There are many variations of the population modeling[20]. The Verhulst model is the classic example to illustrate the periodic doubling and chaotic behavior in dynamical system [19] . The model describs the population growth may be limited by certain factors like population density [20].

2. APPLICATIONS OF LOGISTIC EQUATION

A typical application of the Logistic equation is a common model of population growth. Let $u(t)$represents the population size and t represents the time where the constant $\rho > 0$ defines the growth rate.

Another application of Logistic curve is in medicine, where the Logistic differential equation is used to model the growth of tumors. This application can be considered an extension of the above mentioned use in the framework of ecology.

The solution of Logistic equation is explained the constant population growth rate which not includes the limitation on food supply or spread of diseases [21] . The solution curve of the model is increase exponentially from the multiplication factor up to saturation limit which is maximum carrying capacity [21] ,

$\frac{dN}{dt} = \rho N\left(1 - \frac{N}{K}\right)$ where N is the population with respect to time, ρ is the rate of maximum population growth and K is the carrying capacity. The solution of continuous Logistic equation is in the form of constant growth rate as in formula $N(t) = N_0 e^{\rho t}$ where N_0 is the initial population [22] .

A delay differential equation (DDE) is a differential equation in which the derivative of the function at any time depends on the solution at previous time. Introduction of delay in the model enriches its dynamics and allows a precise description of the real life phenomena. DDEs are proved useful in control systems [23] , lasers, traffic models [24], metal cutting, epidemiology, neuroscience, population dynamics [25] , chemical kinetics [26] , etc. In DDE, one has to provide history of the system over the delay interval $[-\tau, 0]$ as the initial condition. Due to this reason delay systems are infinite dimensional in nature. Because of in infinite dimensionality the DDEs are difficult to analyze analytically and hence the numerical solutions play an important role.

In [27], Mai-Duy, et al. derived an integral collocation approach based on Chebyshev polynomials and used it for solving numerically the bi-harmonic equations. In [28] , Bhrawy and Alofi introduced a new shifted Chebyshev operational matrix of fractional integration of arbitrary order and applied together with spectral tau method for solving linear fractional differential equations. Khader et al.[29] introduced a new approximate formula of the fractional derivative using Legendre series expansion and used it to solve numerically the fractional diffusion equation.

In this article, we extend the previous work and derive some schemes of the integral collocation formulation based on Legendre polynomials. We implement these formulae to solve numerically Riccati, Logistic and delay differential equations.

Our paper is organized as follows: In Section 2, we derive some integration collocation formulations using Legendre series expansion. In Section 3, we give the integral collocation approach for solving Riccati, Logistic and delay differential equations. In Section 4, the paper ends with a brief conclusion and some remarks.

3. INTEGRATION COLLOCATION FORMULATIONS

The well known Legendre polynomials are defined on the interval $[-1,1]$ and can be determined with the aid of the following recurrence formula [30]

$$L_{k+1}(z)=\frac{2k+1}{k+1}zL_k(z)-\frac{k}{k+1}L_{k-1}(z), k=1,2,\cdots,$$

where $L_0(z)=1$ and $L_1(z)=z$. In order to use these polynomials on the interval $[0,1]$, we define the so called shifted Legendre polynomials by introducing the change of variable $z=2x-1$. Let the shifted Legendre polynomials $L_k(2x-1)$ be denoted by $L_k^*(x)$. Then $L_k^*(x)$ can be obtained as follows

$$L_{k+1}^*(x)=\frac{(2k+1)(2x-1)}{(k+1)}L_k^*(x)-\frac{k}{k+1}L_{k-1}^*(x),$$

where $L_0^*(x)=1$ and $L_1^*(x)=2x-1$. The analytic form of the shifted Legendre polynomials $L_k^*(x)$ of degree k is given by

$$L_k^*(x)=\sum_{i=0}^{k}(-1)^{k+i}\frac{(k+i)!}{(k-i)!(i!)^2}x^i, k=2,3,\cdots. \tag{1}$$

Note that $L_k^*(0)=(-1)^k$ and $L_k^*(1)=1$. The orthogonality condition is

$$\int_0^1 L_i^*(x)L_j^*(x)\mathrm{d}x=\begin{cases}\dfrac{1}{2i+1}, & \text{for} \quad i=j,\\ 0, & \text{for} \quad i\neq j.\end{cases}$$

In this work, to consider the differential equation of n-th order, we build the integration collocation method using the truncated Legendre series of degree m to represent the n-th derivative of the unknown function $u(x)$ in the following manner

$$\frac{\mathrm{d}^n u(x)}{\mathrm{d}x^n}\cong\sum_{k=0}^{m}a_k L_k^*(x)=\sum_{k=0}^{m}a_k I_k^{(n)}(x). \tag{2}$$

Using the integration we can obtain the lower-order derivatives and the function itself as follows

$$\frac{\mathrm{d}^{n-1}u(x)}{\mathrm{d}x^{n-1}}\cong\sum_{k=0}^{m}a_k I_k^{(n-1)}(x)+c_1, \tag{3}$$

$$\frac{\mathrm{d}^{n-2}u(x)}{\mathrm{d}x^{n-2}}\cong\sum_{k=0}^{m}a_k I_k^{(n-2)}(x)+c_1x+c_2,\cdots \tag{4}$$

$$\frac{\mathrm{d}u(x)}{\mathrm{d}x}=\sum_{k=0}^{m}a_k I_k^{(1)}(x)+c_1\frac{x^{n-2}}{(n-2)!}+c_2\frac{x^{n-3}}{(n-3)!}+\cdots+c_{n-2}x+c_{n-1}, \tag{5}$$

$$u(x)=\sum_{k=0}^{m}a_k I_k^{(0)}(x)+c_1\frac{x^{n-1}}{(n-1)!}+c_2\frac{x^{n-2}}{(n-2)!}+\cdots+c_{n-1}x+c_n, \tag{6}$$

from (1) and (2) we have

$$I_k^{(n)}(x)=\sum_{i=0}^{k}\frac{(-1)^{k+i}(k+i)!}{(k-i)!(i!)^2}x^i,$$

$$I_k^{(n-1)}(x)=\int I_k^{(n)}(x)\mathrm{d}x=\sum_{i=0}^{k}\frac{(-1)^{k+i}(k+i)!}{(k-i)!(i!)^2(i+1)}x^{i+1},$$

$$I_k^{(n-2)}(x)=\int I_k^{(n-1)}(x)\mathrm{d}x=\sum_{i=0}^{k}\frac{(-1)^{k+i}(k+i)!}{(k-i)!(i!)^2(i+1)(i+2)}x^{i+2},$$

$\cdots$

$$I_k^{(0)}(x)=\int I_k^{(1)}(x)\mathrm{d}x=\sum_{i=0}^{k}\frac{(-1)^{k+i}(k+i)!x^{i+n}}{(k-i)!(i!)^2(i+1)\cdots(i+n-1)(i+n)}. \tag{7}$$

We now collocate Equatuions (2)-(6) at $(m+1)$ points x_p, $p=0,1,\cdots,m$ as

$$\frac{\mathrm{d}^n u(x_p)}{\mathrm{d}x^n}=\Omega^{(n)}\hat{S},\quad \frac{\mathrm{d}^{n-1}u(x_p)}{\mathrm{d}x^{n-1}}=\Omega^{(n-1)}\hat{S},\cdots,$$
$$\frac{\mathrm{d}u(x_p)}{\mathrm{d}x}=\Omega^{(1)}\hat{S},\quad u(x_p)=\Omega^{(0)}\hat{S}, \tag{8}$$

where $\hat{S}=[a_0,a_1,\cdots,a_m,c_1,c_2,\cdots,c_n]^{\mathrm{T}}$, and $\Omega^{(n)},\Omega^{(n-1)},\cdots,\Omega^{(0)}$ are integrated matrices.

4. INTEGRAL COLLOCATION APPROACH FOR SOLVING RICCATI, LOGISTIC AND DELAY DIFFERENTIAL EQUATIONS

In this section, we introduce the integral collocation approach using Legendre expansion for solving the Riccati, Logistic and delay differential equations.

4.1. Model 1: Riccati Differential Equation

$$\frac{\mathrm{d}u(x)}{\mathrm{d}x}+u^2(x)-1=0, \qquad x>0, \tag{9}$$

we also assume an initial condition

$$u(0)=u^0. \tag{10}$$

The exact solution to this problem at $u^0=0$ is

$$u(x)=\frac{\mathrm{e}^{2x}-1}{\mathrm{e}^{2x}+1}.$$

The procedure of the implementation is given by the following steps:

1) Approximate the function $u(x)$ using Formula (6) and its relevant derivatives with $m=5$, as follows

$$\frac{du(x)}{dx} \cong \sum_{k=0}^{5} a_k I_k^{(1)}(x),$$

$$u(x) \cong \sum_{k=0}^{5} a_k I_k^{(0)}(x) + c_1, \tag{11}$$

where $I_k^{(0)}(x)$ is defined in (7) as

$$I_k^{(0)}(x) = \sum_{i=0}^{k} (-1)^{k+i} \frac{(k+i)!}{(k-i)!(i!)^2 (i+1)} x^{i+1},$$

$$I_k^{(1)}(x) = \sum_{i=0}^{k} (-1)^{k+i} \frac{(k+i)!}{(k-i)!(i!)^2} x^{i}.$$

Then the Riccati differential Equation (9) is transformed to the following approximated form

$$\sum_{k=0}^{5} a_k L_k^*(x) + \left(\sum_{k=0}^{5} a_k I_k^{(0)}(x) + c_1 \right)^2 = 1. \tag{12}$$

We now collocate Equation (12) at $(m+1=6)$ points x_p, $p=0,1,2,3,4,5$ as

$$\sum_{k=0}^{5} a_k L_k^*(x_p) + \left(\sum_{k=0}^{5} a_k I_k^{(0)}(x_p) + c_1 \right)^2 = 1. \tag{13}$$

For suitable collocation points we use the roots of shifted Legendre polynomial $L_6^*(x)$.

2) Also, by substituting from the initial condition (10) in (11) we can obtain $(n=1)$ an equation which gives the value of the constant c_1 as follows

$$c_1 = u^0 = 0. \tag{14}$$

Equations (13) and (14) represent a system of non-linear algebraic equations which contains seven equations for the unknowns $a_k, k = 0,1,2,3,4,5$ and c_1.

3) Solve the resulting system using the Newton iteration method to obtain the unknowns $a_k, k = 0,1,2,3,4,5$ as follows

$$a_0 = 0.76160, \quad a_1 = -0.31790, \quad a_2 = -0.0506,$$
$$a_3 = 0.02896, \quad a_4 = -0.00111, \quad a_5 = -0.00112.$$

Therefore, from Formula (12) we can obtain the approximate solution in the form

$$u(x) \cong \sum_{k=0}^{5} a_k I_k^{(0)}(x) + c_1$$
$$s = 0.990x + 0.002x^2 - 0.346x^3 + 0.027x^4 + 0.125x^5 - 0.047x^6.$$

The numerical results of the proposed problem (9) are given in Figure 1 with different values of $m\,(m = 5,7)$ in the interval $[0,1]$ at $u^0 = 0$. From this figure, since the obtained numerical solutions are in excellent agreement with the exact solution, so, we can conclude that the proposed technique is well for solving such class of ODEs.

4.2. Model 2: Logistic Differential Equation

$$\frac{du(x)}{dx} = \rho u(x)\big(1 - u(x)\big), \quad x > 0, \ \rho > 0. \tag{15}$$

We also assume an initial condition

$$u(0) = u_0, \qquad u_0 > 0. \tag{16}$$

The exact solution to this problem is given by

$$u(x) = \frac{u_0}{(1-u_0)e^{-\rho x} + u_0}.$$

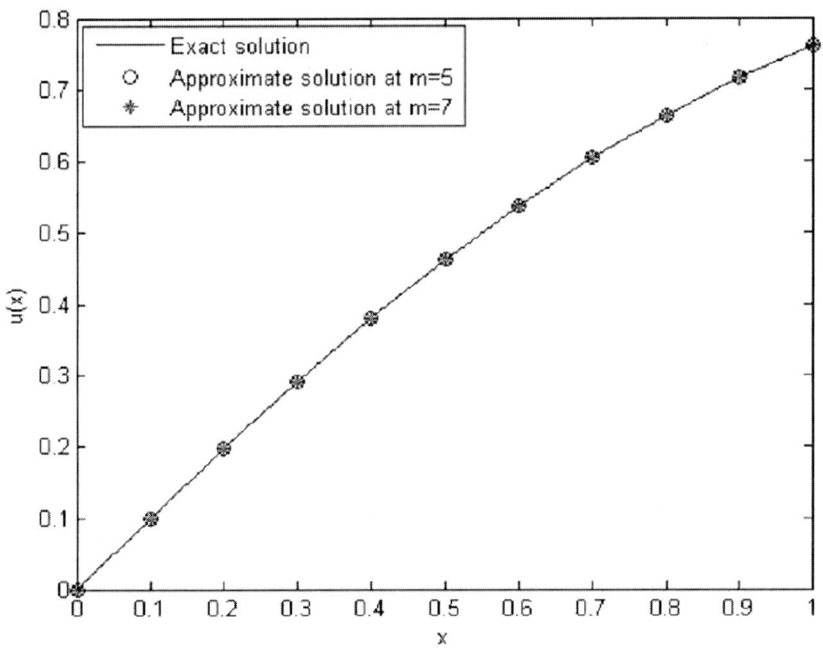

Figure 1. A comparison between the exact solution and the approximate solution at $m = 5$ and $m = 7$.

The existence and the uniqueness of the proposed problem (15) are introduced in details in [31] .

The procedure of the implementation is given by the following steps:

1) Approximate the function $u(x)$ and its relevant derivatives with $m = 5$, using Formula (11). Then the Logistic differential Equation (15) is transformed to the following approximated form

$$\sum_{k=0}^{5} a_k L_k^*(x) - \rho\left(\sum_{k=0}^{5} a_k I_k^{(0)}(x) + c_1\right)$$
$$\left(1 - \left(\sum_{k=0}^{5} a_k I_k^{(0)}(x) + c_1\right)\right) = 0. \tag{17}$$

We now collocate Equation (17) at $(m+1=6)$ points x_p, $p = 0,1,2,3,4,5$ as

$$\sum_{k=0}^{5} a_k L_k^*(x_p) - \rho\left(\sum_{k=0}^{5} a_k I_k^{(0)}(x_p) + c_1\right)\left(1 - \left(\sum_{k=0}^{5} a_k I_k^{(0)}(x_p) + c_1\right)\right) = 0. \tag{18}$$

For suitable collocation points we use roots of shifted Legendre polynomial $L_6^*(x)$ which are

$$x_0 = 0.96623, \quad x_1 = 0.03377,$$
$$x_2 = 0.38069, x_3 = 0.61930,$$
$$x_4 = 0.16931, \quad x_5 = 0.83060. \tag{19}$$

2) Also, by substituting from the initial condition (16) in (11) with $u_0 = 0.85$ we can obtain $(n=1)$ an equation which gives the value of the constant $c_1 = 0.85$.

Equation (18) represents a system of non-linear algebraic equations which contains six equations for the unknowns $a_k, k = 0,1,2,3,4,5$.

Solve the resulting system using the Newton iteration method to obtain the unknowns $a_k, k = 0,1,2,3,4,5$ as follows

$$a_0 = 0.0533, \quad a_1 = -0.0101,$$
$$a_2 = 0.0004, a_3 = 0.00002,$$
$$a_4 = -1.642\times10^{-6}, \ a_5 = 5.398\times10^{-8}. \tag{20}$$

Therefore, from Formula (11) we can obtain the approximate solution in the form

$$u(x) = \sum_{k=0}^{5} a_k I_k^{(0)}(x) + c_1$$
$$= 1 + x + 0.5x^2 + 0.1667x^3 + 0.0417x^4 + 0.0083x^5 + 0.0014x^6 + 0.0002x^7 + 0.00004x^8.$$

The numerical results of the proposed problem (15) are given in Figure 2 with different values of $m (m = 3,5)$ in the interval $[0,1]$. From this figure, since the obtained numerical solutions are in excellent agreement with the exact solution, so, we can conclude that the proposed technique is well for solving such class of ODEs.

4.3. Model 3: Delay Differential Equation

Consider the linear delay differential equation of third-order

$$\frac{d^3u(x)}{dx^3} = -u(x) - u(x-0.3) + e^{-x+0.3}, 0 \le x \le 1, \tag{21}$$

with the initial conditions

$$u(0)=1,\frac{du(0)}{dx}=-1,\frac{d^2u(0)}{dx^2}=1,u(x)=e^{-x}. \tag{22}$$

The exact solution of this model is $u(x)=e^{-x}$.

The procedure of the implementation is given by the following steps:

1) Approximate the function $u(x)$ using Formula (6) and its relevant derivatives with $m = 5$, as follows

$$\begin{aligned}
\frac{d^3u(x)}{dx^3} &\cong \sum_{k=0}^{5} a_k L_k^*(x) = \sum_{k=0}^{3} a_k I_k^{(3)}(x),\\
\frac{d^2u(x)}{dx^2} &\cong \sum_{k=0}^{5} a_k I_k^{(2)}(x) + c_1,\\
\frac{du(x)}{dx} &\cong \sum_{k=0}^{5} a_k I_k^{(1)}(x) + c_1 x + c_2,\\
u(x) &\cong \sum_{k=0}^{5} a_k I_k^{(0)}(x) + c_1\frac{x^2}{2!} + c_2\frac{x}{1!} + c_3,
\end{aligned} \tag{23}$$

where $I_k^{(0)}(x)$, $I_k^{(1)}(x)$ and $I_k^{(x)}(x)$ are defined in (7) as follows

$$\begin{aligned}
I_k^{(0)}(x) &= \sum_{i=0}^{k} \frac{(-1)^{k+i}(k+i)!}{(k-i)!(i!)^2(i+1)(i+2)(i+3)} x^{i+3},\\
I_k^{(1)}(x) &= \sum_{i=0}^{k} \frac{(-1)^{k+i}(k+i)!}{(k-i)!(i!)^2(i+1)(i+2)} x^{i+2},\\
I_k^{(2)}(x) &= \sum_{i=0}^{k} \frac{(-1)^{k+i}(k+i)!}{(k-i)!(i!)^2(i+1)} x^{i+1}.
\end{aligned}$$

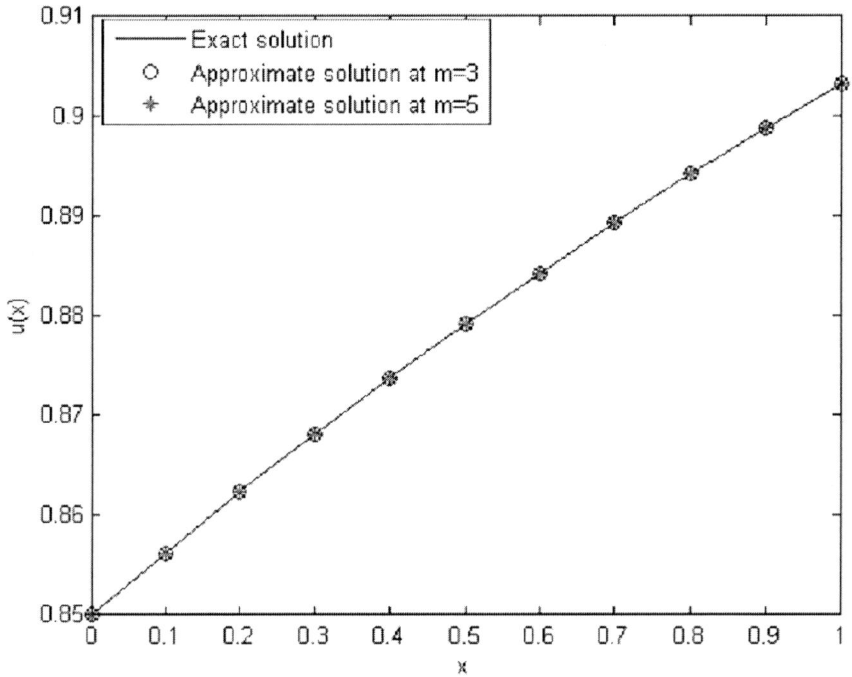

Figure 2. A comparison between the exact solution and the approximate solution at m = 3 and m = 5.

Then the delay differential Equation (21) is transformed to the following approximated form

$$\sum_{k=0}^{5} a_k L_k^*(x) + \left(\sum_{k=0}^{5} a_k I_k^{(0)}(x) + c_1 \frac{x^2}{2!} + c_2 \frac{x}{1!} + c_3 \right)$$
$$+ \left(\sum_{k=0}^{5} a_k I_k^{(0)}(x-0.3) + c_1 \frac{(x-0.3)^2}{2!} + c_2 \frac{(x-0.3)}{1!} + c_3 \right) = \mathrm{e}^{-x+0.3}. \tag{24}$$

We now collocate Equation (24) at $(m+1=6)$ points x_p, $p=0,1,2,3,4,5$ as

$$\sum_{k=0}^{5} a_k L_k^*(x_p) + \left(\sum_{k=0}^{5} a_k I_k^{(0)}(x_p) + c_1 \frac{x_p^2}{2!} + c_2 \frac{x_p}{1!} + c_3 \right) + \left(\sum_{k=0}^{5} a_k I_k^{(0)}(x_p)(x_p - 0.3) + c_1 \frac{(x_p - 0.3)^2}{2!} + c_2 \frac{(x_p - 0.3)}{1!} + c_3 \right) = \mathrm{e}^{-x_p + 0.3}. \tag{25}$$

For suitable collocation points we use the roots of shifted Legendre polynomial $L_6^*(x)$.

2) Also, by substituting from the initial conditions (22) in (23) we can obtain $(n=3)$ equations which give the values of the constants

$$c_1 = 1, \quad c_2 = -1, \quad c_3 = 1. \tag{26}$$

Equations (25) and (26) represent a system of linear algebraic equations which contains nine equations for the unknowns $a_k, k = 0,1,2,3,4,5$ and c_1, c_2, c_3.

3) Solve the resulting system using the conjugate gradient method to obtain the unknowns $a_k, k = 0,1,2,3,4,5$ as follows

$$a_0 = -0.6321, \quad a_1 = 0.3109, \quad a_2 = -0.0515,$$
$$a_3 = 0.0051, \quad a_4 = 0.0004, \quad a_5 = 0.00002. \tag{27}$$

Therefore, from Formula (23) we can obtain the approximate solution in the form

$$u(x) \cong \sum_{k=0}^{5} a_k I_k^{(0)}(x) + c_1 \frac{x^2}{2!} + c_2 \frac{x}{1!} + c_3$$
$$= 1 - x + 0.5x^2 - 0.1667x^3 + 0.0417x^4 - 0.0084x^5 + 0.0014x^6 - 0.0002x^7 + 0.00002x^8.$$

The numerical results of the proposed problem (21) are given in Figure 3 with different values of $m\,(m = 5,7)$ in the interval $[0,1]$. From this figure, since the obtained numerical solutions are in excellent agreement with the exact solution, so, we can conclude that the proposed technique is well for solving such class of ODEs.

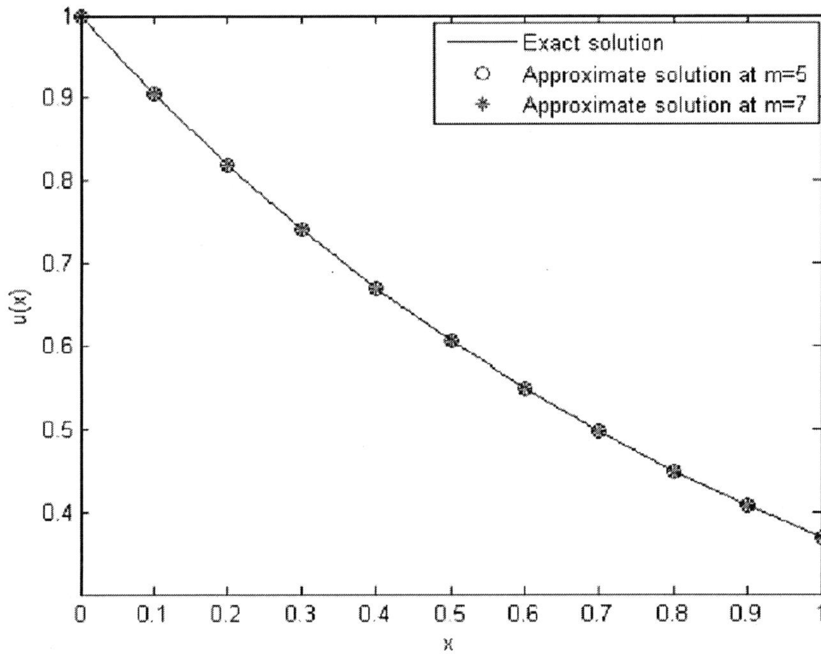

Figure 3. A comparison between the exact solution and the approximate solution at m = 5 and m = 7.

5. CONCLUSION AND REMARKS

In this article, an integral collocation approach based on Legendre polynomials is introduced for solving numerically the Riccati, Logistic and delay differential equations. The properties of the Legendre polynomials are used to reduce the proposed problems to system of algebraic equations which are solved by a suitable numerical method. From the obtained numerical results, we can conclude that this method gives results with an excellent agreement with the exact solution. All numerical results are obtained using Matlab program 8.

REFERENCES

1. Khader, M.M. (2011) On the Numerical Solutions for the Fractional Diffusion Equation. Communications in Nonlinear Science and Numerical Simulation, 16, 2535-2542.

2. Khader, M.M. (2013) Numerical Treatment for Solving the Perturbed Fractional PDEs Using Hybrid Techniques. Journal of Computational Physics, 250, 565-573.

3. Sweilam, N.H., Khader, M.M. and Mahdy, A.M.S. (2013) Numerical Study for the Fractional Differential Equations Generated by Optimization Problem Using Chebyshev Collocation Method and FDM. Applied Mathematics and Information Science, 7, 2013-2020.

4. Khader, M.M. and Hendy, A.S. (2013) A Numerical Technique for Solving Fractional Variational Problems. Mathematical Methods in Applied Sciences, 36, 1281-1289.

5. Sweilam, N.H., Khader, M.M. and Nagy, A.M. (2011) Numerical Solution of Two-Sided Space-Fractional Wave Equation Using Finite Difference Method. Journal of Computional and Applied Mathematics, 235, 2832-2841.

6. Sweilam, N.H., Khader, M.M. and Adel, M. (2012) On the Stability Analysis of Weighted Average Finite Difference Methods for Fractional Wave Equation. Fractional Differential Calculus, 2, 17-75.

7. Sweilam, N.H. and Khader, M.M. (2010) A Chebyshev Pseudo-Spectral Method for Solving Fractional Order Integro-Differential Equations. ANZIAM, 51, 464-475.

8. Sweilam, N.H., Khader, M.M. and Al-Bar, R.F. (2008) Homotopy Perturbation Method for Linear and Nonlinear System of Fractional Integro-Differential Equations. International Journal of Computational Mathematics and Numerical Simulation, 1, 73-87.

9. Sweilam, N.H., Khader, M.M. and Mahdy, A.M.S. (2012) On the Numerical Solution for the Linear Fractional Klein-Gordon Equation Using Legendre Pseudospectral Method. International Journal of Mathematics and Computer Applications Research, 2, 1-10.

10. Khader, M.M., EL-Danaf, T.S. and Hendy, A.S. (2013) A Computational Matrix Method for Solving Systems of High Order Fractional Differential Equations. Applied Mathematical Modelling, 37, 4035-4050.

11. Khader, M.M., Mahdy, A.M.S. and Shehata, M.M. (2014) Approximate Analytical Solution to the Time-Fractional Biological Population Model Equation. Jokull, 64, 378-394.

12. Sweilam, N.H., Khader, M.M. and Mahdy, A.M.S. (2012) Numerical Studies for Solving Fractional-Order Logistic Equation. International Journal of Pure and Applied Mathematics, 78, 1199-1210.

13. Sweilam, N.H., Khader, M.M. and Mahdy, A.M.S. (2012) Numerical Studies for Fractional-Order Logistic Differential Equation with Two Different Delays. Journal of Applied Mathematics, 2012, Article ID: 764894.

14. Reid, W.T. (1972) Riccati Differential Equations Mathematics in Science and Engineering. Academic Press, New York.

15. Lasiecka, I. and Triggiani, R. (1991) Differential and Algebraic Riccati Equations with Application to Boundary/Point Control Problems: Continuous Theory and Approximation Theory. Lecture Notes in Control and Information Sciences, Springer, Berlin.

16. Bahnasawi, A.A., El-Tawil, M.A. and Abdel-Naby, A. (2004) Solving Riccati Differential Equation Using ADM. Applied Mathematics and Computation, 157, 503-514.

17. Tan, Y. and Abbasbandy, S. (2008) Homotopy Analysis Method for Quadratic Riccati Differential Equation. Communications in Nonlinear Science and Numerical Simulation, 13, 539-546.

18. Cushing, J.M. (1998) An Introduction to Structured Population Dynamics, Society for Industrial and Applied Mathematics.

19. Alligood, K.T., Sauer, T.D. and Yorke, J.A. (1996) An Introduction to Dynamical Systems. Springer, Berlin.

20. Ausloos, M. The Logistic Map and the Route to Chaos: From the Beginnings to Modern Applications XVI. 411, 2006.

21. Pastijn, H. (2006) Chaotic Growth with the Logistic Model of P.-F. Verhulst, Understanding Complex Systems. The Logistic Map and the Route to Chaos. 3-11.

22. Suansook, Y. and Paithoonwattanakij, K. (2009) Dynamic of Logistic Model at Fractional Order. IEEE International Symposium on Industrial Electronics.

23. Fridman, E., Fridman, L. and Shustin, E. (2000) Steady Modes in Relay Control Systems with Time Delay and Periodic Disturbances. Journal of Dynamic Systems, Measurement, and Control, 122, 732-737.

24. Davis, C.L. (2002) Modification of the Optimal Velocity Traffic Model to Include Delay due to Driver Reaction Time. Physica A, 319, 557-567.

25. Kuang, Y. (1993) Delay Differential Equations with Applications in Population Biology. Academic Press, Boston, San Diego, New York.

26. Epstein, I. and Luo, Y. (1991) Differential Delay Equations in Chemical Kinetics: Nonlinear Models: The Cross-Shaped Phase Diagram and the Originator. Journal of Chemical Physics, 95, 244-254.

27. Mai-Duy, N., See, H. and Tran-Cong, T. (2009) A Spectral Collocation Technique Based on Integrated Chebyshev Polynomials for Biharmonic Problems in Irregular Domains. Applied Mathematical Modelling, 33, 284-299.

28. Bhrawy, A.H. and Alofi, A.S. (2013) The Operational Matrix of Fractional Integration for Shifted Chebyshev Polynomials. Applied Mathematics Letters, 26, 26-31.

29. Khader, M.M., Sweilam, N.H. and Mahdy, A.M.S. (2011) An Efficient Numerical Method for Solving the Fractional Diffusion Equation. Journal of Applied Mathematics and Bioinformatics, 1, 1-12.

30. Bell, W.W. (1968) Special Functions for Scientists and Engineers. Great Britain, Butler and Tanner Ltd, Frome and London.

31. El-Sayed, A.M.A., El-Mesiry, A.E.M. and El-Saka, H.A.A. (2007) On the Fractional-Order Logistic Equation. Applied Mathematics Letters, 20, 817-823.

CHAPTER 9

The Adomian Decomposition Method and the Differential Transform Method for Numerical Solution of Multi-Pantograph Delay Differential Equations

Musa Cakir, Derya Arslan

Department of Mathematics, Faculty of Science,
University of Yuzuncu Yil, Van, Turkey

ABSTRACT

In this paper, the Adomian Decomposition Method (ADM) and the Differential Transform Method (DTM) are applied to solve the multi-pantograph delay equations. The sufficient conditions are given to assure the convergence of these methods. Several examples are presented to demonstrate the efficiency and reliability of the ADM and the DTM; numerical results are discussed, compared with exact solution. The results of the ADM and the DTM show its better performance than others. These methods give the desired accurate results only in a few terms and in a series form of the solution. The approach is simple and effective. These methods are used to solve many linear and nonlinear problems and reduce the size of computational work.

KEYWORDS

Multi-Pantograph Delay Differential Equations, Adomian Decomposition Method (ADM), Differential Transform Method (DTM), Convergence of Adomian Decomposition Method

1. INTRODUCTION

Pantograph is a device located on the electriclocomotive. The first time, electric locomotive was made in America in 1851. It was commissioned in 1895. Mathematical model of pantograph was first developed by Taylor and Ockendon (1971) [1] . Pantograph equations belong to a special class of functional-differential equations with proportional delays and arise in many applications such as, astrophysics, nonlinear dynamical systems, probabi- lity theory on algebraic structures, electro dynamics, quantum mechanics and cell growth, number theory, mixing problems, population models, etc.

In recent years, the multi-pantograph delay differential equations were studied by many authors. For examples, Li and Liu [2] applied the Runge-Kutta methods to the multi-pantograph delay equation. Evans and Raslan [3] used the Adomian decomposition method for solving the delay differential equation. Keskin et al. [4] applied the differential transform method to obtain the approximate solution. Sezer and Dascioglu [5] developed and applied the Taylor method to the generalized pantograph equation with retarded case or advanced case. Brunner [1] used the collocation methods for pantograph-type Volterra functional equations with multiple delays. Yu [6] applied the variational iteration method to the multi-pantograph delay equation. Sezer et al. [7] worked approxi- mate solution of multi-pantograph equation with variable coefficients. Geng, F. Z. and Qian, S. P. [8] worked the Reprociding Kernel Medhod to Solving Singularly Perturbed Multi-Pantograph Delay Equations. Cherruault, Y., Adomian, G., Abbaoui, K. and Rach, R. [9] worked on Convergence of Decomposition Method. Ismail et al. [10] gave the numerical solutions of the Korteweg-De-Vries (KDV) and modified Korteweg-De-Vries Equations. El-Safty et al. [11] studied on the 3-h step spline function approximation to the solution of delay dynamic system. Saeed and Rahman [12] established the differential transform method to solve systems of linear or non-linear delay differential equation.

A numerical method based on the Adomian Decomposition Method (ADM) which has been used from the 1970s to the 1990s by George Adomian [13] [14] . The differential transform method (DTM) has been successfully developed by Zhou (1986) in electric circuit analysis. DTM has been used to solve linear and nonlinear differential equations [15] .

ADM and DTM have been shown to solve effectively, easily and accurately a large class of linear and nonlinear, ordinary, partial, deterministic or stochastic differential equations with approximate solutions which converge rapidly to accurate solutions [3] [4] [13] [14] . The basic motivation of this work is to apply the ADM and DTM to the DDE. It is well known now in the literature that this algorithm provides the solution in a rapidly convergent series [3] [4] . ADM and DTM are very effective and convenient for solving multi-pantograph equations [3] [4] .

This study is presented as follows: In second section, we start by presenting ADM and DTM to solve multi-pantograph delay differential equations. In third section, we continue to the presentation of the convergence of ADM with Theorem 3.1 and

Definition 3.2. In fourth section, these methods are shown and compared by four examples by taking various values for t and error evaluation is made. Also, we have plotted the graphs for numerical solutions of ADM and DTM and exact solution.

We examined that multi-pantograph delay differential equations are solved by several methods. Thus, we wanted to show up that may be more efficient, simpler and reliable the solution treatment of the ADM for multi- pantograph delay differential equations. The results show that the ADM is more powerful method than other methods for multi-pantograph delay differential equations.

2. ANALYSIS OF ADOMIAN DECOMPOSITION METHOD AND THE DIFFERENTIAL TRANSFORM

In this paper, we consider the following multi-pantograph equations [16] ,

$$y'(t)=a(t)y(t)+\sum_{i=1}^{k}b_i(t)y(q_it)+f(t),\ 0\le t<T \tag{1}$$

$$y(t)=y_0, \tag{2}$$

$a(t),b_i(t),f(t)$ are analytical functions, $0<q_i<1, i=1,\cdots,k$.

Using the ADM, the differential operator L is given by

$$L(.)=\frac{d^n}{dt^n}(.) \tag{3}$$

The inverse operator L^{-1}, this is n-fold integral operator defined by

$$L^{-1}(.) = \int_0^t (.)_{n\text{-times}} \, \mathrm{d}t \tag{4}$$

operating with L^{-1} on Equation (1), it then follows

$$Ly + Ry + Ny = f \tag{5}$$

where the method defines Ny the nonlinear term by the Adomian polynomials [3] [13] [14]

$$Ny = \sum_{n=0}^{\infty} A_n,$$

A_n are Adomian polynomials that can be generated for all forms of nonlinearity as

$$A_n = \frac{1}{n!}\frac{\mathrm{d}^n}{\mathrm{d}\alpha^n}\left[N\left(\sum_{i=1}^{n}\alpha^i y_i\right)\right]_{\alpha=0},$$

where $i = 0,1,2,3,\cdots$

$$A_0 = f(y_0)$$

$$A_1 = y_1\left(\frac{\mathrm{d}}{\mathrm{d}y_0}\right)f(y_0)$$

$$A_2 = y_2\left(\frac{\mathrm{d}}{\mathrm{d}y_0}\right)f(y_0) + \left(\frac{y_1^2}{2!}\right)\left(\frac{\mathrm{d}^2}{\mathrm{d}y^2}\right)f(y_0)$$

$$A_3 = y_3\left(\frac{\mathrm{d}}{\mathrm{d}y_0}\right)f(y_0) + y_1 y_2\left(\frac{\mathrm{d}^2}{\mathrm{d}y_0^2}\right)f(y_0) + \left(\frac{y_1^3}{3!}\right)\left(\frac{\mathrm{d}^3}{\mathrm{d}y_0^3}\right)f(y_0).$$

Operating with L^{-1} on Equation (5), it then follows

$$L^{-1}Ly + L^{-1}Ry + L^{-1}Ny = L^{-1}f \tag{6}$$

$$y = y(0) + L^{-1}f - L^{-1}Ry - L^{-1}Ny \tag{7}$$

$$\sum_{n=0}^{\infty} y_n = y_0 + L^{-1}f - L^{-1}R\sum_{n=0}^{\infty} y_n - L^{-1}\sum_{n=0}^{\infty} A_n \tag{8}$$

to determine the components

$$y_n(t), n \geq 0.$$

First, we identify the zero component $y_0(t)$ by all terms that arise from the boundary conditions at t = 0 and from integrating the source term if it exists. Second, the remaining components of $y(t)$ can be determined in a way such that each component is determined by using the preceding components [3] [13] [14]

$$y_0 = y(0) + L^{-1}f, \tag{9}$$

and Equation (8) gives for $n = 1, 2, 3, \cdots$ in other words, the method introduces the recursive relation

$$\begin{aligned} y_1 &= -L^{-1}Ry_0 - L^{-1}A_0 \\ y_2 &= -L^{-1}Ry_1 - L^{-1}A_1 \\ y_3 &= -L^{-1}Ry_2 - L^{-1}A_2 \\ &\vdots \\ y_{n+1} &= -L^{-1}Ry_n - L^{-1}A_n \end{aligned} \tag{10}$$

The Adomian decomposition method assumes that the unknown function $y(t)$ can be expressed by an infinite series of the form [3] [13] [14]

$$y(t)=\sum_{n=0}^{\infty}y_n=y_0+y_1+y_2+\cdots$$

so that the components $y_n(t)$ will be determined recursively [3] [13] [14]

2.1 Differential Transform Method

Differential transform of function $y(x)$ is defined as follows [17] ,

$$Y(k)=\frac{1}{k!}\left[\frac{\mathrm{d}^k y(x)}{\mathrm{d}x^k}\right]_{x=0} \tag{11}$$

In Equation (11), $y(x)$ is original function and $Y(k)$ is the transformed function, which is called the T- function. Differential inverse transform of $Y(k)$ is defined as

$$Y(k)=\sum_{k=0}^{\infty}x^k Y(k) \tag{12}$$

From Equations (12) and (11), we obtain

$$y(x)=\sum_{k=0}^{\infty}\frac{x^k}{k!}\left[\frac{\mathrm{d}^k y(x)}{\mathrm{d}x^k}\right]_{x=0} \tag{13}$$

Equation (13) implies that concept of differential transform is derived from Taylor series expansion, but the method does not evaluate the derivatives symbolically.

In actual applications, the function $y(x)$ is expressed by the a finite series and Equation (12) can be written as

$$y(x)=\sum_{k=0}^{m}x^kY(k) \tag{14}$$

Equation (13) implies that is

$$y(x)=\sum_{k=m+1}^{\infty}x^kY(k)$$

is negligibly small. In fact, m is decided by the convergence of natural frequency in this study.

The following theorems that can be deduced from Equations (11) and (12) are given below, see [17] [18] .

Theorem 1 If $y(t)=g(t)+h(t)$, then $Y(k)=G(k)+H(k)$.

Theorem 2 If $y(t)=cg(t)$, then $Y(k)=cG(k)$.

Theorem 3 If $y(t)=\frac{\mathrm{d}^k g(t)}{\mathrm{d}^k}$, then $Y(k)=\frac{(k+n)!}{k!}G(k+n)$.

Theorem 4 If $y(t)=g(t+a)$, then $Y(k)=\sum_{m=k}^{N}\binom{h}{k}a^{h-k}G(m)$, for $N\to\infty$.

Theorem 5 If $y(t)=g\left(\frac{t}{a}\right)a\geq 1$, then

$$Y(k)=\sum_{h=k}^{N}(-1)^{h-k}\frac{(a-1)^{h-k}}{a^h}t_0^{h-k}\binom{h}{k}a^{h-k}G(h),\text{ for } N\to\infty.$$

3. THE CONVERGENCE OF ADM

The Adomian Decomposition Method is equivalent to the sequence defined as follows [19] [20]

$$S_n = y_1 + y_2 + y_3 + \cdots + y_n,$$

by using the iterative scheme

$$S_0 = 0,$$
$$S_{n+1} = N(y_0 + S_n) \tag{15}$$

and related to the functional equation

$$S = N(y_0 + S).$$

The numerical solution of Equation (15) was used fixed-point theorem by Cherruault [19] [20] .

Theorem 3.1 Let N be an operator from Hilbert space H in to H and y be the exact solution of functional equation. $\sum_{i=0}^{\infty} y_i$, which is obtained by ADM iterative scheme, converges to y when $\exists 0 \le \alpha < 1$, $\|y_{k+1}\| \le \alpha \|y_k\|$,

$\forall k \in N \cup \{0\}$.

Proof See [19] [20] .

Definition 3.2 For every $i \in N \cup \{0\}$ we define

$$\alpha_i = \begin{cases} \dfrac{\|y_{i+1}\|}{\|y_i\|}, & \|y_i\| \ne 0, \\ 0, & \|y_i\| = 0, \end{cases}$$

Corollary 3.3 In Theorem 4.1, $\sum_{i=0}^{\infty} y_i$ converges to exact solution y, when

$0 \le \alpha_i < 1, i = 1,2,3,\cdots$

4. NUMERICAL EXAMPLES

In this section, four experiments of multi-pantograph delay differential equations are given to illustrate the efficiency of the ADM and the DTM. The examples are computed using Maple 15. Results obtained by the methods are compared with the exact solution of each example and found to be good agreement with each other. The absolute errors in tables are given at selected points.

Example 4.1 Consider the following linear multi-pantograph delay equation of the first-order [16]

$$y'(t) = -\frac{5}{6}y(t) + 4y\left(\frac{t}{2}\right) + 9y\left(\frac{t}{3}\right) + t^2 - 1,\ 0 < t \le 1,$$
$$y(0) = 1,$$

which has the exact solution,

$$y(t) = 1 + \frac{67}{6}t + \frac{1675}{72}t^2 + \frac{12157}{1296}t^3.$$

Using the ADM, we get according to Equations (3)-(10), we obtain recursive formula for $k = 0,1,2,3,\cdots$,

$$y_{k+1} = L^{-1}\left(-\frac{5}{6}y_k(t) + 4y_k\left(\frac{t}{2}\right) + 9y_k\left(\frac{t}{3}\right)\right)$$

$$y_0 = 1 + \frac{1}{3}t^3 - t$$

$$y_1 = \frac{73}{6}t - \frac{25}{12}t^2$$

$$y_2 = \frac{1825}{72}t^2 - \frac{175}{216}t^3$$

$$y_3 = \frac{12775}{1296}t^3$$

Thus, we obtain

$$y_{\text{ADM}}(t) = \sum_{n=0}^{\infty} y_n = 1 + \frac{67}{6}t + \frac{1675}{72}t^2 + \frac{12157}{1296}t^3 + 0 + 0 + \cdots$$

The solution by DTM method:

By using Theorems of DTM, we have following recurrence relation:

$$Y(k+1) = \frac{Y(k)}{k+1}\left[-\frac{5}{6} + \frac{4}{2^k} + \frac{9}{3^k}\right] + \delta(k-2) - \delta(k),\ k \ge 1,\ Y(0) = 1$$

Utilizing the recurrence relation, we find

$$Y(1) = \frac{67}{6},\ Y(2) = \frac{1675}{72},\ Y(3) = \frac{12157}{1296},\ Y(4) = 0,\ Y(5) = 0, \cdots$$

Finally, the differential inverse transform of $Y(k)$ gives

$$y(t) = \sum_{k=0}^{\infty} Y(k) t^k$$

we obtain the following series solution

$$y_{\text{DTM}}(t)=\sum_{k=0}^{\infty}Y(k)t^k=1+\frac{67}{6}t+\frac{1675}{72}t^2+\frac{12157}{1296}t^3+0+0+\cdots$$

the closed form of above solution is

$$y(t)=1+\frac{67}{6}t+\frac{1675}{72}t^2+\frac{12157}{1296}t^3$$

which is exactly the same as the exact solution.

The obtained results (ADM and DTM) are exactly the same with the one found by exact solution. It is clear from Table 1 and Figure 1 that the three results not only give rapidly convergent series but also accurately compute the solutions.

Using our methods, we choose 6 points on [0, 1] respectively. The numerical results are given in the following Table 1.

Example 4.2 Solve the following nonlinear pantograph delay equation of first-order [3]

$$y'(t)=1-2y^2\left(\frac{t}{2}\right),\quad 0\le t\le 1$$

$$y(0)=0,$$

which has the exact solution, $y(t)=\sin t.$

The solution by ADM method:

By applying the ADM, according to Equations (3)-(10), we obtain

$$Ly=1-2y^2\left(\frac{t}{2}\right)$$

$$L^{-1}Ly=L^{-1}\left(-2y^2\left(\frac{t}{2}\right)\right)+L^{-1}(1)$$

$$y(t) = L^{-1}\left(-2y^2\left(\frac{t}{2}\right)\right) + t$$

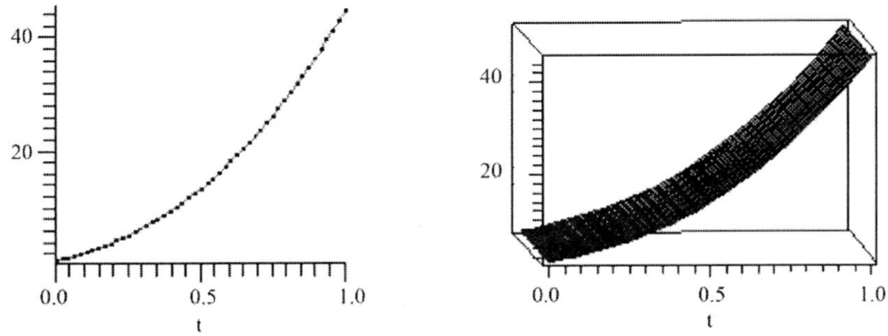

Figure 1. The obtained solution of the multi-pantograph delay equation [ADM solution (red), DTM solution (black), EXACT solution (blue)].

We have solved this problem using the proposed method. Recursive formula and the sequence of approximate solution are obtained as follows:

$$y_{k+1}(t) = -L^{-1}\left(2y_k^2\left(\frac{t}{2}\right)\right)$$

$$k = 0, y_1(t) = -\frac{t^3}{6}$$

$$k = 1, y_2(t) = \frac{t^5}{120}, \quad k = 2, y_3(t) = -\frac{t^7}{5040}$$

Table 1. Comparison of exact solution, the (ADM) and the (DTM) of y(t).

t	Solution		Error
	y_{ADM}/y_{DTM}	y_{exact}	
0	1	1	0
0.2	4.238932099	4.238932099	0
0.4	9.789234568	9.789234568	0
0.6	18.10116667	18.10116667	0
0.8	29.62498766	29.62498766	0
1	44.8109568	44.8109568	0

thus, we obtain:

$$y_{\mathrm{ADM}}(t)=\sum_{n=0}^{\infty}y_n=t-\frac{t^3}{3!}+\frac{t^5}{5!}-\frac{t^7}{7!}+\cdots$$

Using to convergence of ADM's method,

$$\alpha_i=\left\|\frac{y_{i+1}}{y_i}\right\|, 0\le\alpha_i<1, i=0,1,2,3,\cdots.$$

$$\alpha_0=0.5000000000<1$$
$$\alpha_1=0.02380952381<1$$
$$\alpha_2=0.01302729713<1$$
$$\vdots$$

Here, the values of α_i are less than one and hence ADM is convergent.

The solution by DTM method:

By using Theorems of DTM, we have following recurrence relation:

$$(k+1)Y(k+1)=\delta(k)-2\sum_{r=0}^{k}Y(r)Y(k-r),\ k\ge 1,\ Y(0)=1$$

Utilizing the recurrence relation, we find

$$Y(1)=1, Y(2)=0, Y(3)=-\frac{1}{3!}, Y(4)=0, Y(5)=\frac{1}{5!},\cdots$$

Finally, the differential inverse transform of $Y(k)$ gives

$$y(t)=\sum_{k=0}^{\infty}Y(k)t^k$$

we obtain the following series solution

$$y_{\text{DTM}}(t) = t - \frac{t^3}{3!} + \frac{t^5}{5!} - \frac{t^7}{7!} + \cdots$$

The obtained results (ADM and DTM) are exactly the same with each other. Increasing the approximation order up to the absolute differences between the numerical solutions are calculated for and comparisons have been made with known results as reported in Table 2 and Figure 2.

Example 4.3 Consider the following linear multi-pantograph delay equation of the first-order [21] .

$$y'(t) = -3y(t) + 0.4y(0.4t) + y(0.1t),\ 0 < t \leq 1,$$
$$y(0) = 2.127,$$

The solution by ADM method:

By applying the ADM, according to Equations (3)-(10), we obtain

$$Ly = -3y(t) + 0.4y(0.4t) + y(0.1t)$$

$$L^{-1}Ly = L^{-1}\left(-3y(t) + 0.4y(0.4t) + u(0.1t)\right)$$

$$y(t) = 2.127 - L^{-1}\left(3y(t) - y(0.4t) - y(0.1t)\right)$$

We have solved this problem using the proposed method. Recursive formula and the sequence of approximate solution are obtained as follows:

$$y_{k+1} = L^{-1}\left(-3y_k(t) + 0.4y_k(0.4t) + y_k(0.1t)\right)$$

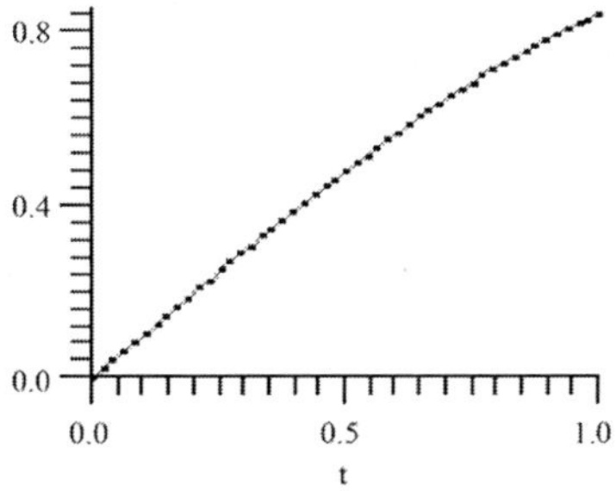

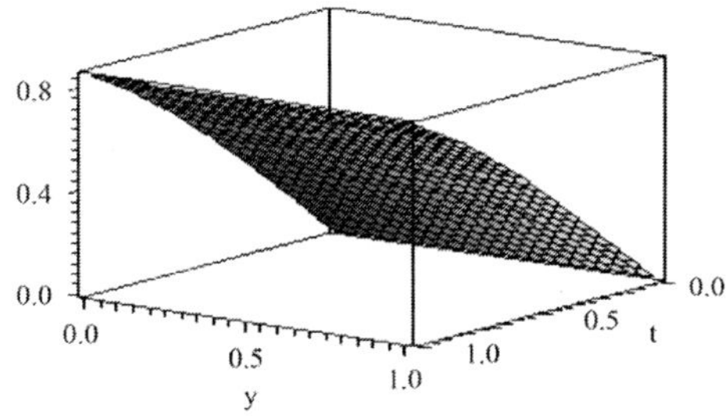

Figure 2. The obtained solution of the multi-pantograph delay equation [ADM solution (red), DTM solution (blue), EXACT solution (black)].

$$y_0 = 2.127,$$

$$k = 0,\ y_1 = -3.403200000t$$

$$k = 1,\ y_2 = 4.662384000t^2$$

$$k = 2,\ y_3 = -4.547378527t^3$$

$$k = 3,\ y_4 = 3.380293828t^4$$

Thus, we obtain:

$$y_{\text{ADM}}(t) = 2.127 - 3.403200000t + 4.662384000t^2 - 4.547378527t^3 + 3.380293828t^4 - \cdots$$

The solution by DTM method:

By using Theorems of DTM, we have following recurrence relation:

$$Y(k+1) = \frac{Y(k)}{k+1}\left[-3 + \left(\frac{2}{5}\right)^{k+1} + \frac{1}{10^k}\right],\ k \geq 1,\ Y(0) = 2.127$$

Table 2. Comparison of the ADM, the DTM and the EXACT solution.

t	Solution		
	y_{ADM}	y_{DTM}	y_{EXACT}
0	0	0	0
0.2	0.198669331	0.198669	0.198669333
0.4	0.389418342	0.389418	0.389418342
0.6	0.564642446	0.564642	0.564642446
0.8	0.717355723	0.717356	0.717355723
1	0.841468254	0.841471	0.841468254

Using the recurrence relation, we find

$$Y(1) = -3.403200000, Y(2) = 4.662384000, Y(3) = -4.547378528,$$
$$Y(4) = 3.380293829, Y(5) = -2.021185850, \cdots$$

Finally, the differential inverse transform of $Y(k)$ gives

$$y(t) = \sum_{k=0}^{\infty} Y(k) t^k$$

we obtain the following series solution

$$y(t) = 2.127 - 3.403200000t + 4.662384000t^2 - 4.547378528t^3 + 3.380293829t^4 - 2.021185850t^5 + \cdots$$

The obtained results (ADM and DTM) are exactly the same with the one found by exact solution. It is clear from Table 3 and Figure 3 that the two results not only give rapidly convergent series but also accurately compute the solutions.

Example 4.4 Consider the following linear multi-pantograph delay equation of the first-order [21] .

$$y'(t) = -y(t) - 0.8y(0.5t) + 0.1y(0.25t), 0 < t \leq 1,$$
$$y(0) = 1,$$

The solution by ADM method:

By applying the ADM, according to Equations (3)-(10), we obtain

$$Ly = -y(t) - 0.8y(0.5t) + 0.1y(0.25t)$$

$$Ly = L^{-1}\left(-y(t) + 0.8y(0.5t) + 0.1u(0.25t)\right)$$

$$y(t) = 1 - L^{-1}\left(y(t) + 0.8y(0.5t) - 0.1y(0.25t)\right)$$

We have solved this problem using the proposed method. Recursive formula and the sequence of approximate solution are obtained as follows:

$$y_{k+1} = L^{-1}\left(-y_k(t) + 0.8y_k(0.5t) + 0.1y_k(0.25t)\right)$$

$$y(0) = 1,\ y_0 = 1,$$

$$k = 0,\ y_1 = -1.700000000t$$

$$k = 1,\ y_2 = 1.68750000t^2$$

$$k = 2,\ y_3 = -0.4650651040t^3$$

$$k = 3,\ y_4 = 0.1277112376t^4$$

Thus, we obtain:

$$y_{\text{ADM}}(t) = \sum_{n=0}^{\infty} y_n = 1 - 1.700000000t + 1.68750000t^2 - 0.4650651040t^3 + \cdots$$

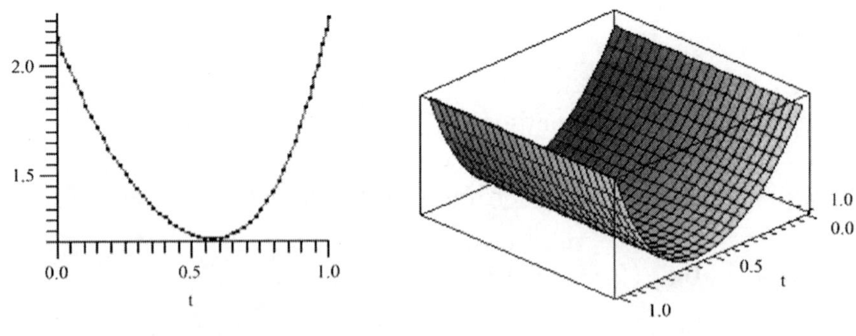

Figure 3. The obtained solution of the multi-pantograph delay equation [ADM solution (red), DTM solution (blue)].

Table 3. Comparison of the ADM solution y_{ADM} with the solution by DTM is illustrated.

t	Solution	
	y_{ADM}	y_{DTM}
0	2.127	2.127
0.2	1.601884802	1.601884802
0.4	1.307204736	1.307204736
0.6	1.219390558	1.219390558
0.8	1.444676306	1.444676306
1	2.219099301	2.219099301

Using to convergence of ADM's method,

$$\alpha_i = \left\| \frac{y_{i+1}}{y_i} \right\|, 0 \le \alpha_i < 1, i = 0,1,2,3,\cdots.$$

$$\alpha_0 = 0.9926470588 < 1$$
$$\alpha_1 = 0.2755941357 < 1$$
$$\alpha_2 = 0.2746093751 < 1$$
$$\vdots$$

according to the obtained results ADM is convergent to the exact solution.

The solution by DTM method:

By using Theorems of DTM, we have following recurrence relation,

$$Y(k+1) = \frac{Y(k)}{k+1}\left[-1 + \frac{0.8}{2^k} + \frac{0.1}{4^k}\right], k \ge 1, Y(0) = 1$$

Utilizing the recurrence relation, we find

$$Y(1) = -1.7, Y(2) = 1.168750000t, Y(3) = -0.4650651042, Y(4) = 0.127712376, \cdots$$

Finally, the differential inverse transform of $Y(k)$ gives

$$y(t) = \sum_{k=0}^{\infty} Y(k) t^k$$

we obtain the following series solution

$$y(t) = 1 - 1.7t + 1.168750000t^2 - 0.4650651042t^3 + 0.127712376t^4 - \cdots$$

The obtained results (ADM and DTM) are exactly the same with the one found by exact solution. It is clear from Table 4 and Figure

4 that the two results not only give rapidly convergent series but also accurately compute the solutions.

5. CONCLUSION

It has been the aim of this paper to show that it appears natural to approximate the solution of multi-pantograph delay differential equation by ADM and DTM. We obtain the high approximate solutions or the exact solutions within a few iterations. It is concluded from figures and tables that the successive approximations methods are an accurate and efficient method to solve multi-pantograph delay differential equations. Some numerical examples have been provided to illustrate that the present method is effective in accuracy and convergence speed. In a word, the ADM and DTM show that the techniques are reliable, powerful and promising methods for linear and nonlinear problems.

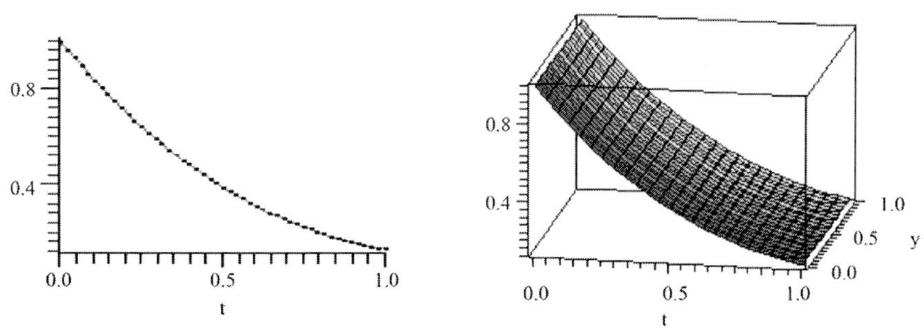

Figure 4. The obtained solution of the multi-pantograph delay equation [ADM solution (red), DTM solution (blue)].

Table 4. Comparison of the ADM solution y_{ADM} with the solution by DTM is illustrated.

t	Solution	
	y_{ADM}	y_{DTM}
0.0	1	1
0.2	0.703233817	0.703233817
0.4	0.480505241	0.480505241
0.6	0.316847313	0.316847313
0.8	0.202197189	0.202197189
1.0	0.131396133	0.131396133

REFERENCES

1. Ockendon, J.R. and Taylor, A.B. (1971) The Dynamics of a Current Collection System for an Electric Locomotive. Proceedings of the Royal Society of London A, 322, 447-468.

2. Li, D. and Liu, M. (2005) Runge-Kutta Methods for the Multi-pantograph Delay Equation. Applied Mathematics and Computation, 163, 383-395.

3. Evans, D.J. and Raslan, K.R. (2005) The Adomian Decomposition Method for Solving Delay Differential Equation. International Journal of Computer Mathematics, 82, 49-54.

4. Keskin, Y., at al. (2007) Approximate Solutions of Generalized Pantograph Equations by the Differential Transform Method. International Journal of Nonlinear Sciences and Numerical, 8, 159-164.

5. Sezer, M. and Dascioglu, A.A. (2007) A Taylor Method for Numerical Solution of Generalized Pantograph Equations with Linear Functional Argument. Journal of Computational and Applied Mathematics, 200, 217-225.

6. Yu, Z. (2008) Variational ?teration Method for Solving the Multi-Pantograph Delay Equation. Physics Letters A, 372, 6475-6479.

7. Sezer, M., Yalcinbas, S. and Sahin, N. (2008) Approximate Solution of Multi-Pantograph Equation with Variable Coefficients. Journal of Computational and Applied Mathematics, 214, 406-416.

8. Geng, F.Z. and Qian, S.P. (2014) Solving Singularly Perturbed Multi-Pantograph Delay Equations Based on the Reprociding Kernel Medhod. Abstract and Applied Analysis, 2014, 6 p.

9. Cherruault, Y., Adomian, G., Abbaoui, K. and Rach, R. (1995) Further Remarks on Convergence of Decomposition Method. International Journal of Bio-Medical Computing, 38, 89-93.

10. Ismail, H.N., Raslan, K.R. and Salem, G.S. (2004) Solitary Wave Solutions for the General KdV Equation by Adomian Decomposition Method. Applied Mathematics and Computation, 154, 17-29.

11. El-Safty, A., Salim, M.S. and El-Khatib, M.A. (2003) Convergent of the Spline Functions for Delay Dynamic System. International Journal of Computer Mathematics, 80, 509-518.

12. Rostam, K., Saeed, B. and Rahman, M. (2011) Differential Transform Method for Solving System of Delay Differential Equation. Australian Journal of Basic and Applied Sciences, 5, 201-206.

13. Adomian, G. (1994) Solving Frontier Problems of Physics: The Decomposition Method. Kluwer Academic Publishers, Boston.

14. Adomian, G. and Rach, R. (1993) Analytic Solution of Nonlinear Boundary-Value Problems in Several Dimensions by Decomposition. Journal of Mathematical Analysis and Applications, 174, 118-137.

15. Zhou, J.K. (1986) Differential Transform and Its Application for Electrical Circuits. Huazhong University Press, Wuhan.

16. Feng, X. (2013) An Analytic Study on the Multi-Pantograph Delay Equations with Variable Coefficients. Bulletin Mathematique de la society des Sciences mathématiques de Roumanie Tome, 56, 205-215. http://ssmr.ro/bulletin/pdf/56-2/articol_7.pdf

17. Ayaz, F. (2004) Applications of Differential Transform Method to Differential-Algebraic Equations. Applied Mathematics and Computation, 152, 649-657.

18. Kurnaz, A. and Oturanc, G. (2005) The Differential Transform Approximation for the System Ordinary Differential Equations. International Journal of Computer Mathematics, 82, 709-719.

19. Cherruault, Y. (1989) Convergence of Adomian's Method. Kybernetes, 18, 31-38.

20. Hosseini, M.M. and Nasabzadeh, H. (2006) On the Convergence of Adomian Decomposition Method. Applied Mathematics and Computation, 182, 536-543.

21. Liu, M.Z. and Li, D. (2004) Properties of Analytic Solution and Numerical Solution of Multi-Pantograph Equation. Applied Mathematics and Computation, 155, 853-871.

CHAPTER 10

On a fractional-order delay Mackey-Glass equation

Ahmed MA El-Sayed 1*, Sanaa M Salman 2 and Naemaa A Elabd 3

1 Department of Mathematics, Faculty of Science, Alexandria University, Alexandria, Egypt.

2 Faculty of Education, Alexandria University, Alexandria, Egypt.

3 Faculty of Science, Alexandria University, Alexandria, Egypt.

ABSTRACT

In this paper, a fractional-order Mackey-Glass equation with constant delay is considered. The local stability of the fixed points is analyzed. Moreover, a discretization process is applied to convert the fractional-order delay equation to its discrete analog. A numerical simulation including Lyapunov exponent, phase diagrams, bifurcation, and chaos is carried out using Matlab to ensure theoretical results and to reveal more complex dynamics of the equation after discretization.

KEYWORDS

fractional-order delay Mackey-Glass equations, fixed points, local stability, discretization, Lyapunov exponent, bifurcation, chaos

1. INTRODUCTION

Delay differential equations (DDEs) arise in the mathematical description of systems whose time evolution depends explicitly on a past state of the system, as for example in the case of delayed feedback. Neural systems [1], respiration regulation [2], agricultural commodity markets [3], nonlinear optics, and neutrophil populations in the blood [2] are but a few systems in which delayed feedback leads naturally to a description in terms of a delay differential equation. We will restrict our attention to systems modeled by evolutionary delay equations that can be expressed in the form

$$x'(t) = f\left(x(t), x(t-\tau)\right), \quad x(t) \in \Re^n, t \geq 0. \tag{1.1}$$

Here the 'state' of the system at time t is x(t) whose rate of change depends explicitly, via the function f, on the past state x(t−τ) where τ is a fixed time delay. More general delay equations might be considered: multiple time delays, variable time delays, continuously distributed delays, and higher derivatives all arise in applications and lead to more complicated evolution equations. Nevertheless, equations of the form (1.1) constitute a sufficiently broad class of systems to be of practical importance, and they will provide adequate fodder for the types of problems we wish to consider.

DDEs arise in many areas of mathematical modeling: for example, population dynamics (taking into account the gestation times), infectious diseases (accounting for the incubation periods), physiological and pharmaceutical kinetics (modeling, for example, the body's reaction to CO_2, *etc.* in circulating blood), chemical kinetics (such as mixing reactants), the navigational control of ships and aircraft (with, respectively, large and short lags), and more general control problems (see for example [4, 5, 6]).

On the other hand, fractional calculus is a generalization of classical differentiation and integration to arbitrary (non-integer) order [7, 8, 9]. Many mathematicians and applied researchers have tried to model real processes using fractional calculus [10, 11, 12, 13, 14, 15, 16]. In recent years differential equations with fractional-order have attracted many researchers because of their applications in many areas of science and engineering. Analytical and numerical techniques have been implemented to study such equations. The fractional calculus has allowed the operations of integration and differentiation to be applied for any fractional order [17, 18, 19, 20, 21].

We recall the basic definitions (Caputo) and properties of fractional-order differentiation and integration.

Definition 1

The fractional integral of order $\beta \in \mathbb{R}^+$ of the function f(t), t>0 is defined by

$$I_a^\beta f(t) = \int_0^t \frac{(t-s)^{\beta-1}}{\Gamma(\beta)} f(s)\,ds,$$

and the fractional derivative of order $\alpha \in (n-1, n)$ of $f(t)$, $t > 0$ is defined by

$$D_a^\alpha f(t) = I^{n-\alpha} D^n f(t), \quad D = \frac{d}{dt}.$$

In addition, the following results are the main features in fractional calculus. Let $\mathbb{R}^+, \alpha \in (0,1)$:

- $I_a^\beta : L^1 \to L^1$, and if $f(x) \in L^1$, then $I_a^\gamma I_a^\beta f(x) = I_a^{\gamma+\beta} f(x)$.
- $\lim_{\beta \to n} I_a^\beta f(x) = I_a^n f(x)$ uniformly on $[a,b]$, $n = 1,2,3,\ldots$, where $I_a^1 f(x) = \int_a^x f(s)\,ds$.
- $\lim_{\beta \to 0} I_a^\beta f(x) = f(x)$ weakly.
- If $f(x)$ is absolutely continuous on $[a,b]$, then $\lim_{\alpha \to 1} D_a^\alpha f(x) = \frac{df(x)}{dx}$.

The Mackey-Glass equation is a nonlinear time delay differential equation, which was proposed as a model of hematopoiesis, given by

$$\frac{dx}{dt} = \frac{\rho x_\tau}{1 + x_\tau^c} - \gamma x, \tag{1.2}$$

where γ, c, ρ, τ are real parameters, and xτxτ represents the value of the variable x at time (t−τ). Depending on the values of the parameters, this equation displays a range of periodic and chaotic dynamics.

In this work, we will show that considering a fractional-order derivative with delay in equation (1.2) will exhibit more complex and richer dynamics.

Consider the fractional-order delay Mackey-Glass equation given in the form

$$D^{\alpha}x(t) = \frac{\rho x(t-\tau)}{1+x(t-\tau)^{c}} - x(t-\tau), \quad t \in (0,T], \tag{1.3}$$

with the initial condition

$$x(0) = x_0, \tag{1.4}$$

where $\alpha \in (0,1]$, $\rho \in R^{+}$, and $c > 0$. In equation (1.3), we consider delay in the last term.

2. DISCRETIZATION PROCESS

In this part, we apply the discretization process represented in [22, 23], and [24] for discretizing the delay fractional-order Mackey Glass equation with piecewise constant arguments given by

$$D^{\alpha}x(t) = \frac{\rho x([\frac{t}{r}]r-\tau)}{1+x^{c}([\frac{t}{r}]r-\tau)} - x\left(\left[\frac{t}{r}\right]r-\tau\right), \tag{2.1}$$

with initial condition (1.4).

The steps of the discretization process are as follows.

Let $t \in [0,r)$, then $\frac{t}{r} \in [0,1)$. That is,

$$D^{\alpha}x(t) = \frac{\rho x_0(-\tau)}{1+x_0^{c}(-\tau)} - x_0(-\tau), \tag{2.2}$$

and the solution of (2.2) is given by

$$x(t) = x_0 + I^{\alpha}\left(\frac{\rho x_0(-\tau)}{1+x_0^c(-\tau)} - x_0(-\tau)\right)$$

$$= x_0 + \left(\frac{\rho x_0(-\tau)}{1+x_0^c(-\tau)} - x_0(-\tau)\right)\int_0^t \frac{(t-s)^{\alpha-1}}{\Gamma(\alpha)}ds$$

$$= x_0 + \frac{t^{\alpha}}{\Gamma(1+\alpha)}\left(\frac{\rho x_0(-\tau)}{1+x_0^c(-\tau)} - x_0(-\tau)\right).$$

Let $t \in [r, 2r)$, then $\frac{t}{r} \in [1, 2)$. That is,

$$D^{\alpha}x(t) = \frac{\rho x_1(r-\tau)}{1+x_1^c(r-\tau)} - x_1(r-\tau), \quad t \in [r, 2r), \tag{2.3}$$

and the solution of (2.3) is given by

$$x_2(t) = x_1(r) + I_r^{\alpha}\left(\frac{\rho x_1(r-\tau)}{1+x_1^c(r-\tau)} - x_1(r-\tau)\right)$$

$$= x_1(r) + \left(\frac{\rho x_1(r-\tau)}{1+x_1^c(r-\tau)} - x_1(r-\tau)\right)\int_r^t \frac{(t-s)^{\alpha-1}}{\Gamma(\alpha)}ds$$

$$= x_1(r) + \frac{(t-r)^{\alpha}}{\Gamma(1+\alpha)}\left(\frac{\rho x_1(r-\tau)}{1+x_1^c(r-\tau)} - x_1(r-\tau)\right).$$

Repeating the process we can easily get

$$x(t) = x_n(nr) + \frac{(t-nr)^{\alpha}}{\Gamma(1+\alpha)}\left(\frac{\rho x_n(nr-\tau)}{1+x_n^c(nr-\tau)} - x_n(nr-\tau)\right), \quad t \in \left[nr, (n+1)r\right).$$

Let t→(n+1)r, we obtain the discretization

$$x_{n+1}(n+1) = x_n + \frac{(r)^{\alpha}}{\Gamma(1+\alpha)}\left(\frac{\rho x_n(r-\tau)}{1+x_n^c(r-\tau)} - x_n(r-\tau)\right). \tag{2.4}$$

It is worth to pay attention here that Euler's discretization method is an approximation for the derivative while the predictor-corrector method is an approximation for the integral. However, our proposed discretization method here is an approximation for the right-hand side of the system under consideration as is pretty clear from (2.4). Moreover, we have noticed that when α→1α→1, the discretization will be Euler's discretization.

In the following, we will discuss two cases of the delay: Case I: τ=r, and Case II: τ=2r

3. CASE I: T=RT=R

In this case we have a second-order difference equation given by

$$x_{n+1} = x_n + \frac{r^\alpha}{\Gamma(1+\alpha)}\left(\frac{\rho x_{n-1}}{1+x_{n-1}^c} - x_{n-1}\right). \tag{3.1}$$

Existence and stability of fixed points

To find the fixed points of system (3.1), we first split it into two first-order difference equations as follows:

$$x_{n+1} = x_n + \frac{r^\alpha}{\Gamma(1+\alpha)}\left(\frac{\rho y_n}{1+y_n^c} - y_n\right),$$

$$y_{n+1} = x_n. \tag{3.2}$$

- For all values of the parameter ρ, system (3.2) has one fixed point, namely, fix1=(0,0)

- For ρ>1, we have an additional fixed point, which is $fix_2 = (\sqrt[c]{\rho - 1}, \sqrt[c]{\rho - 1})$.

In order to study the local stability of these fixed points, we need the moduli of the eigenvalues of the Jacobian matrix evaluated at each of the fixed points [25]. The Jacobian matrix of system (3.2) evaluated at any fixed point (x,y) is obtained by

$$J = \begin{pmatrix} 1 & \frac{r^{\alpha}}{\Gamma(1+\alpha)}\left(\frac{\rho+\rho y^{c}(1-c)}{(1+y^{c})^{2}} - 1\right) \\ 1 & 0 \end{pmatrix}.$$

The eigenvalues associated to the Jacobian matrix are

$$\lambda_{1,2} = 0.5 \pm 0.5\sqrt{1 + 4R(s-1)},$$

where

$$R = \frac{r^{\alpha}}{\Gamma(1+\alpha)}, \qquad S = \frac{\rho + \rho y^{c}(1-c)}{(1+y^{c})^{2}}.$$

The fixed points fix1fix1, fix2 of the system equation (3.1) are stable if $|\lambda i|<1|$, i=1,2. In order to study the qualitative behavior of the solution of system (3.2) we rely on the Jury criteria given generally by

1. $F := 1 + T + D > 0,$
2. $TC := 1 - T + D > 0,$
3. $H := 1 - D > 0,$

where the trace and determinant of the Jacobian matrix are given, respectively, by

T: $\mathrm{Tr}(J) = J_{11} + J_{22}$,

D: $\mathrm{Det}(J) = J_{11}J_{22} - J_{12}J_{21}$.

Proposition 1

The fixed point fix1fix1 *is locally asymptotically stable if* ρ<(1+2/R), *and losses stability via a flip bifurcation when*ρ>1 *and via a Neimark-Sacker bifurcation when* ρ>rα−Γ(1+α)rα.

Proof

The Jacobian matrix at the first fixed point fix1fix1 is obtained by

$$J = \begin{pmatrix} 1 & \frac{r^{\alpha}}{\Gamma(1+\alpha)}(\rho - 1) \\ 1 & 0 \end{pmatrix},$$

which has two eigenvalues,

$$\lambda_{1,2} = 0.5 \pm 0.5\sqrt{1 + 4R(\rho - 1)}.$$

According to the Jury criteria [26, 27], where $T = 1, D = \frac{r^{\alpha}(1-\rho)}{\Gamma(1+\alpha)}$'), the first condition is always satisfied, while the second and third may be violated. That is, the fixed point fix1fix1 loses stability via a flip bifurcation when ρ>1, and via a Neimark-Sacker bifurcation when ρ> $\frac{r^{\alpha} - \Gamma(1+\alpha)}{r^{\alpha}}$. □

Proposition 2

The fixed point fix2fix2 *of system* (3.2) *is stable if*

$\rho < \frac{cR}{2+cR}$, *and it loses stability via a pitchfork bifurcation if* $\rho > \frac{cR}{2+cR}$, *via a flip bifurcation if* $\rho>1$, *and via a Neimark-Sacker bifurcation if* $\rho < \frac{cR}{cR-1}$.

Proof

Calculating the Jacobian matrix at the second fixed point fix2fix2 of system (3.2) we obtain

$$J = \begin{pmatrix} 1 & \frac{r^{\alpha} c(1-\rho)}{\rho\Gamma(1+\alpha)} \\ 1 & 0 \end{pmatrix},$$

which has two eigenvalues,

$$\lambda_{1,2} = 0.5 \pm 0.5\sqrt{1 + \frac{4Rc(1-\rho)}{\rho}},$$

where the trace and determinant of J(fix2) are given, respectively, by

$$T = 1, \qquad D = -Rc\left(\frac{1-\rho}{\rho}\right).$$

According to the Jury criteria, the three conditions may be all violated. That is, fix2 loses stability via a pitchfork bifurcation if $\rho > \frac{cR}{2+cR}$, via a flip bifurcation if $\rho>1$, and via a Neimark-Sacker bifurcation if $\rho < \frac{cR}{cR-1}$.

4. CASE II: T=2R

In this section, we take the delay to be τ=2r in equation (1.3). Applying the discretization process we end up with a system of third-order difference equations given by

$$x_{n+1} = x_n + \frac{r^\alpha}{\Gamma(1+\alpha)}\left(\frac{\rho x_{n-2}}{1+x_{n-2}^c} - x_{n-2}\right). \tag{4.1}$$

To study the fixed points of system (4.1) we first split it into three first-order difference equations as follows:

$$x_{n+1} = x_n + \frac{r^\alpha}{\Gamma(1+\alpha)}\left(\frac{\rho y_n}{1+z_n^c} - z_n\right),$$

$$y_{n+1} = x_n,$$

$$z_{n+1} = y_n. \tag{4.2}$$

In the following, we study the local stability of the fixed points of the system (4.2).

Existence and stability of fixed points

System (4.2) has the following fixed points:

- For all parameter values, there is only one fixed point fixx1=(0,0,0)
- For ρ>1ρ>1, there is an additional fixed point fix $x_2 = (\sqrt[c]{\rho-1}, \sqrt[c]{\rho-1}, \sqrt[c]{\rho-1})$.

By considering a Jacobian matrix for one of these fixed points and calculating their eigenvalues, we can investigate the stability of each fixed point based on the roots of the system characteristic equation. The Jacobian matrix is given by

$$J = \begin{pmatrix} 1 & 0 & R(S-1) \\ 1 & 0 & 0 \\ 0 & 1 & 0 \end{pmatrix},$$

where $R = \frac{r^{\alpha}}{\Gamma(1+\alpha)}$, $S = \frac{\rho+\rho z^{c}(1-c)}{(1+z^{c})^{2}}$.

Linearizing the system (4.2) about fixx1fixx1 yields the following characteristic equation:

$$P(\lambda) = \lambda^3 - \lambda^2 - R(\rho - 1). \tag{4.3}$$

Let

$$a_0 = 1, \qquad a_1 = -1, \qquad a_2 = 0, \qquad a_3 = -R(\rho - 1).$$

From the Jury test, $P(1) > 0, P(-1) < 0, \text{and } a_3 < 1, |b_3| > b_1, c_3 > |c_2|,$, where $b_2 = a_1 - a_3 a_2, b_1 = a_2 - a_3 a_1, c_3 = b_3^2 - b_1^2, \text{and } c_2 = b_3 b_2 - b_1 b_2,$, and $c_2 = b_3 b_2 - b_1 b_2,$, then the roots of P(λ) satisfy λ<1 and thus fix1 is asymptotically stable.

The first condition gives ρ<1, while the second condition gives

$$\rho > \frac{r^{\alpha} - 2\Gamma(\alpha + 1)}{r^{\alpha}}.$$

The remaining conditions give the following inequalities:

- $a_3 < 1 \Rightarrow \rho > 1 - \frac{1}{R},$
- $|b_3| > b_1 \Rightarrow (1 - (R - \rho R)^2) > (R - \rho R),$
- $c_3 > |c_2| \Rightarrow (1 - (R - \rho R)^2)^2 - (R - \rho R)^2 > (R - \rho R)^2 - 1 + (R - \rho R).$

Linearizing the system (4.2) about fixx2fixx2 yields the following characteristic equation:

$$P(\lambda) = \lambda^3 - \lambda^2 - \frac{cR(1-\rho)}{\rho}. \tag{4.4}$$

Let

$$a_0 = 1, \qquad a_1 = -1, \qquad a_2 = 0, \qquad a_3 = -\frac{cR(1-\rho)}{\rho}.$$

From the Jury test, if $P(1) > 0$, $P(-1) < 0$, and $a_3 < 1$, $|b_3| > b_1$, $c_3 > |c_2|$, where $b_3 = 1 - a_3^2$, , $b_2 = a_1 - a_3 a_2$, $b_1 = a_2 - a_3 a_1$, $c_3 = b_3^2 - b_1^2$, and $c_2 = b_3 b_2 - b_1 b_2$, , then the roots of P(λ)P(λ) satisfy λ<1λ<1and thus fix1fix1 is asymptotically stable.

We are going to check these conditions at fixx2fixx2:

- $p(1) > 0 \Rightarrow \rho > 1$,
- $p(-1) < 0 \Rightarrow \rho < \frac{cr^\alpha}{cr^\alpha - 2\Gamma(\alpha+1)}$,
- $a_3 < 1 \Rightarrow \rho < \frac{cr^\alpha}{cr^\alpha - \Gamma(\alpha+1)}$,
- $|b_3| > b_1 \Rightarrow (1 + \frac{c^2R^2}{\rho^2}(1-\rho)^2) > \frac{cR(\rho-1)}{\rho}$.

Thus, any condition may be violated resulting in instability of fixx2.

5. NUMERICAL SIMULATION

In this section, a numerical simulation is carried out with the aid of Matlab to illustrate our theoretical results and to reveal the more complex dynamics of equation (1.2) in the two cases τ=r and τ=2r. In all numerical simulations, we take c=6, and r=0.5. First of all, let us consider system (3.2). Indeed, if one is interested in determining whether a dynamical system is chaotic or not, often just a few of the largest Lyapunov characteristic exponents (LCEs) may provide the answer. This actually is so because a positive LCE is a good indicator for chaos. Since for non-chaotic systems all LCEs are non-positive, the presence of a positive LCE has often been used to help determine if a system is chaotic or not. In this paper, we compute the LCEs via the Householder QR-based methods described in [28]. For system (3.2), we get when α=0.95 c=6 and r=0.5 , LCE1=0.3397, and LCE2=0.185as shown in Figure 1. We vary the parameter ρ and fix the other parameters, c, r, and α. Bifurcation diagrams of system (3.2) are also shown in Figure 1 for different values of the fractional-order parameterα. If we consider α=0.95, it is shown that the fixed point fix1(0,0) is stable if ρ<1, and at ρ>1 it losses stability via a flip bifurcation. Afterwards, a stable periodic solution of period 2 appears, then the periodic solution of period 4 becomes unstable, and a periodic solution of period 8 appears and chaos happens. Figure 2 shows the different phase plane for system (3.2) for α=0.95. For ρ=1.55, Figure 2(a) shows an invariant closed curve bifurcating from fix1(0,0), while for ρ=3.6, Figure 2(b) shows a chaotic attractor. Now we vary the parameter $\alpha$from 0 to 1 and fix ρ to plot the bifurcation diagram for system (3.2) as a function of α as shown in Figure 3.

Next, we turn to the second case, when τ=2r . Figure 4 shows the bifurcation diagram for system (4.2) as a function of ρ. If ρ=0.95, the figure shows that the fixed point fixx1fixx1 becomes unstable when ρ>1ρ>1 as it loses stability via a flip bifurcation. Then the stable period-2 orbit appears at ρ=1.2, which in turn loses stability; then chaos appears. Finally, Figure 5 shows the phase plane for system (4.2) for α=0.95. Figure 5(a) shows a stable fixed point fixx1fixx1 for ρ=0.6, Figure 5(b) shows a double scroll for ρ=1.2, and Figure 5(c), (d) shows chaotic attractors for ρ=2 and ρ=2.7, respectively.

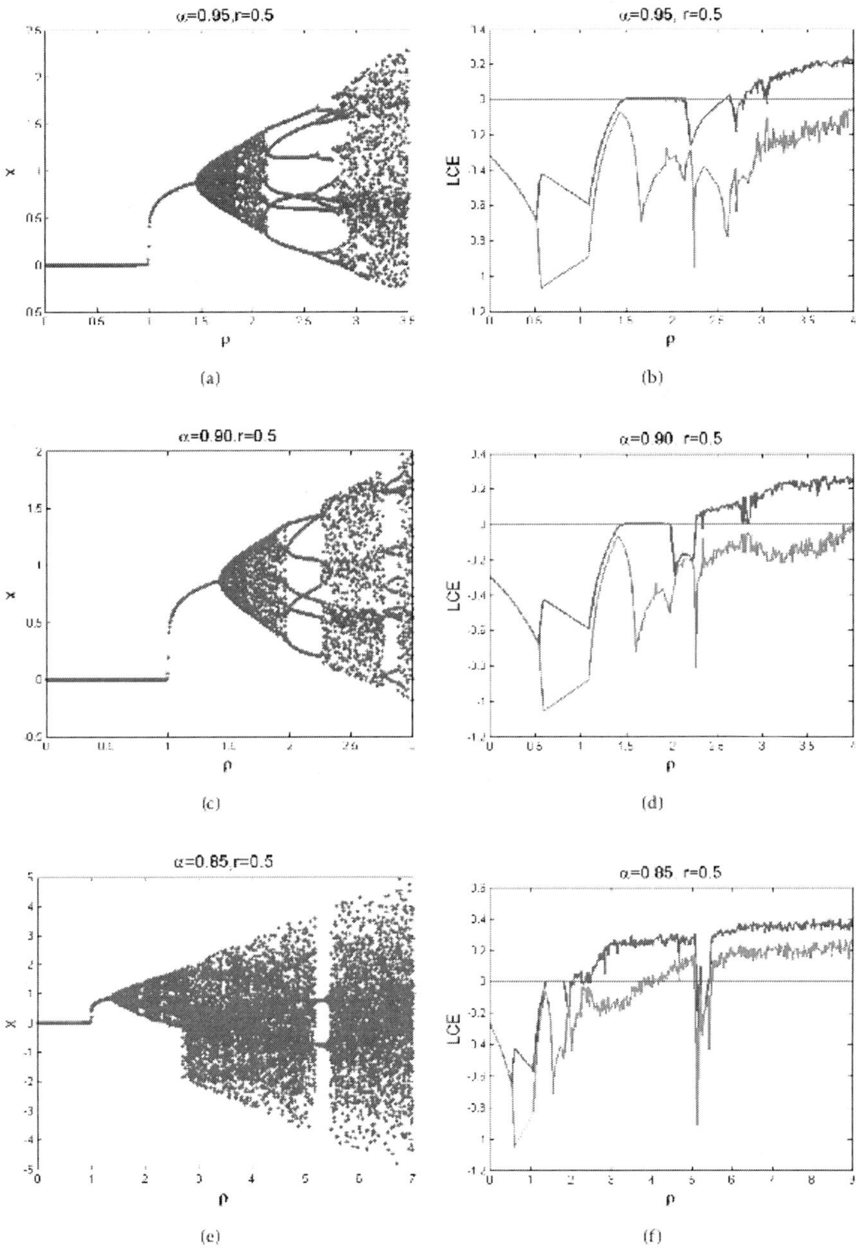

Figure 1. Bifurcation diagram and Lyapunov exponent for system (3.2) with different values of the fractional-order **parameter** $\boldsymbol{\alpha}$ **,** c=6 **and** r=0.5r=0.5 **.**

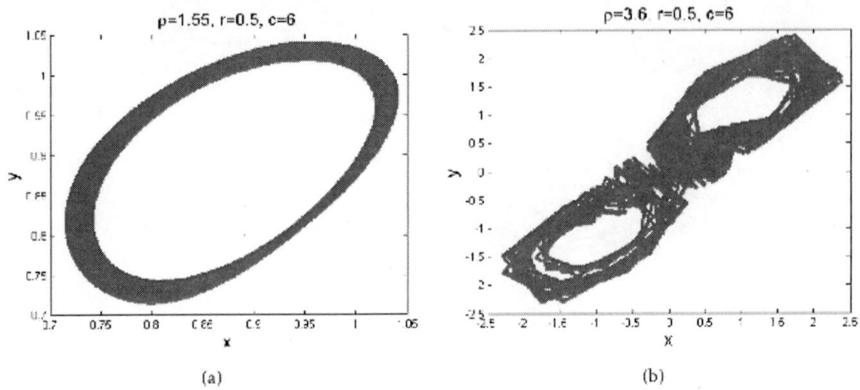

Figure 2. Phase plane for system (3.2) $\alpha = 0.95$, $c = 6$ and $r = 0.5$.

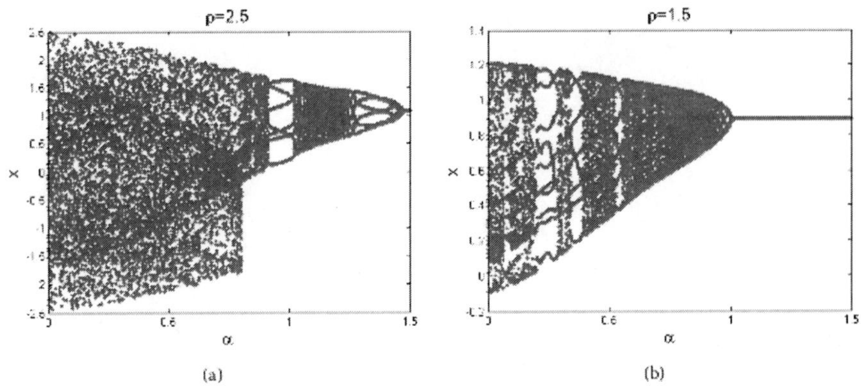

Figure 3. Bifurcation diagrams for system (3.2) as a function of α.

6. CONCLUSION

In this paper, the dynamic behavior of a fractional-order delay Mackey-Glass equation is investigated after applying a discretization process to it. We have considered two different cases for the delay τ, the first is when τ=r, and the second is when τ=2r, where r is the discretization parameter. Stability of the fixed points and local bifurcations of fixed points of the discretized systems in the two cases was are analyzed. A numerical simulation was carried out to ensure our theoretical analysis and to reveal the more complex dynamics of the system.

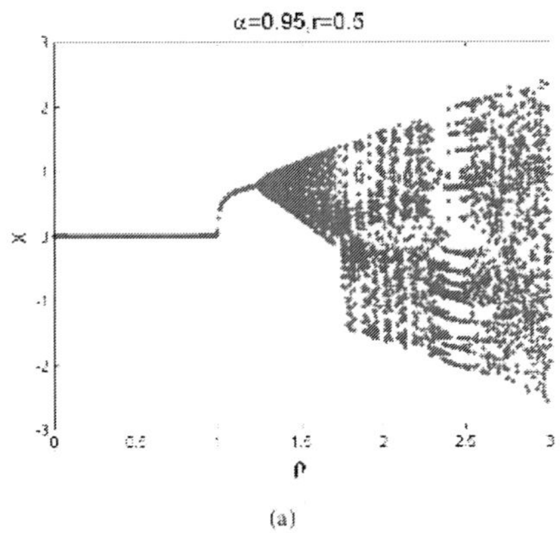

Figure 4. Bifurcation diagram and chaos for system (4.2) as a function of $\boldsymbol{\rho}$.

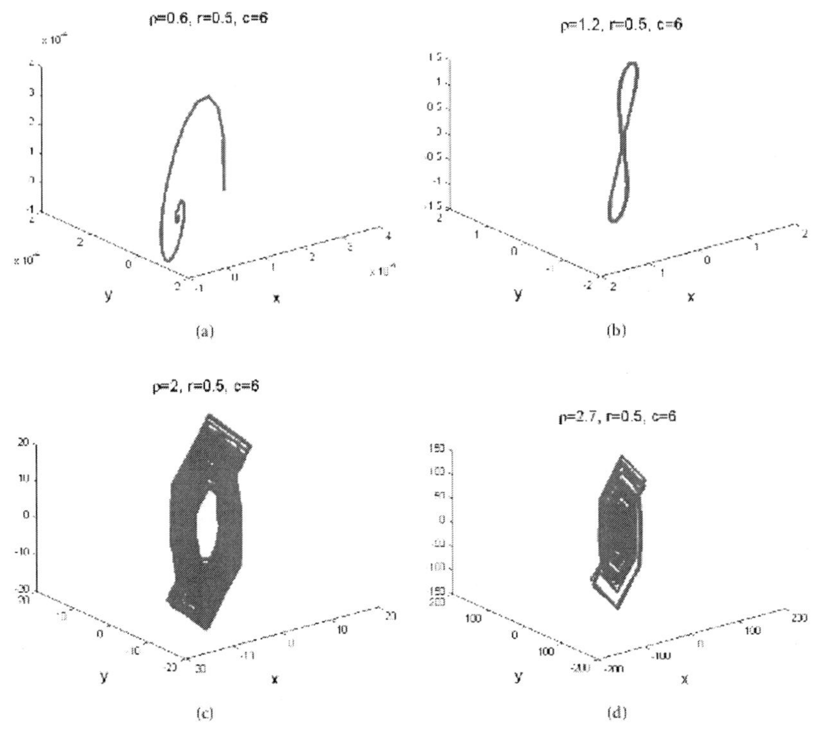

Figure 5. Phase plane for system (3.1) with $\alpha = 0.95, c = 6,$ and $r = 0.5.$

ACKNOWLEDGEMENTS

The authors would like to thank the referees of this manuscript for their valuable comments and suggestions.

COMPETING INTERESTS

The authors declare that they have no competing interests.

AUTHORS' CONTRIBUTIONS

The authors declare that the study was realized in collaboration with the same responsibility. All authors read and approved the final manuscript.

REFERENCES

1. An der Heiden, U: Delays in physiological systems. J. Math. Biol. **8**, 345-364 (1979)
2. Glass, L, Mackey, MC: Pathological conditions resulting from instabilities in physiological control systems. Ann. N.Y. Acad. Sci. **316**, 214-235 (1979)
3. Mackey, MC: Commodity price fluctuations: price dependent delays and non- linearities as explanatory factors. J. Econ. Theory **48**(2), 497-509 (1989)

4. Baleanu, D, Magin, RL, Bhalekar, S, Daftardar-Gejji, V: Chaos in the fractional order nonlinear Bloch equation with delay. Commun. Nonlinear Sci. Numer. Simul. **25**(1-3), 41-49 (2015)

5. Wu, G-C, Baleanu, D: Discrete chaos in fractional delayed logistic maps. Nonlinear Dyn. **80**(4), 1697-1703 (2015)

6. Jarad, F, Abdeljawad, T, Baleanu, D: Higher order fractional variational optimal control problems with delayed arguments. Appl. Math. Comput. **218**(18), 9234-9240 (2012)

7. Kilbas, AA, Srivastava, HM, Trujillo, JJ: Theory and Applications of Fractional Differential Equations. Elsevier, Amsterdam (2006)

8. Podlubny, I: Fractional Differential Equations. Mathematics in Science and Engineering, vol. 198. Academic Press, San Diego (1999)

9. Miller, KS, Ross, B: An Introduction to the Fractional Calculus and Fractional Differential Equation. Wiley, New York (1993)

10. Yang, XJ, Srivastava, HM, Cattani, C: Local fractional homotopy perturbation method for solving fractal partial differential equations arising in mathematical physics. Rom. Rep. Phys. **67**(3), 752-761 (2015)

11. Yang, XJ, Srivastava, HM: An asymptotic perturbation solution for a linear oscillator of free damped vibrations in fractal medium described by local fractional derivatives. Commun. Nonlinear Sci. Numer. Simul. **29**(1), 499-504 (2015)

12. Yang, XJ, Baleanu, D, Srivastava, HM: Local fractional similarity solution for the diffusion equation defined on Cantor sets. Appl. Math. Lett. **47**, 54-60 (2015)

13. Wang, J, Ye, Y, Pan, X, Gao, X, Zhuang, C: Fractional zero-phase filtering based on the Riemann-Liouville integral. Signal Process. **98**(5), 150-157 (2014)

14. Wang, J, Ye, Y, Gao, Y, Gao, X, Qian, S: Fractional compound integral with application to ECG signal denoising. Circuits Syst. Signal Process. **34**, 1915-1930 (2015)

15. Wang, J, Ye, Y, Pan, X, Gao, X: Parallel-type fractional zero-phase filtering for ECG signal denoising. Biomed. Signal Process. Control **18**, 36-41 (2015)

16. Wang, J, Ye, Y, Gao, X: Fractional 90° phase-shift filtering based on the double-sided Grunwald-Letnikov differintegrator. IET Signal Process. **9**(4), 328-334 (2015)

17. Bhrawy, AH, Alhamed, YA, Baleanu, D, Al-Zahrani, AA: New spectral techniques for systems of fractional differential equations using fractional-order generalized Laguerre orthogonal functions. Fract. Calc. Appl. Anal. **17**, 1137-1157 (2014)

18. Hafez, RM, Ezz-Eldien, SS, Bhrawy, AH, Ahmed, EA, Baleanu, D: A Jacobi Gauss-Lobatto and Gauss-Radau collocation algorithm for solving fractional Fokker-Planck equations. Nonlinear Dyn. **82**(3), 1431-1440 (2015)

19. Bhrawy, AH, Hafez, RM, Alzahrani, E, Baleanu, D, Alzahrani, AA: Generalized Laguerre-Gauss-Radau scheme for the first order hyperbolic equations in a semi-infinite domain. Rom. J. Phys. **60**, 918-934 (2015)

20. Bhrawy, AH, Abdelkawy, MA, Alzahrani, AA, Baleanu, D, Alzahrani, E: A Chebyshev-Laguerre Gauss-Radau collocation scheme for solving time fractional sub-diffusion equation on a semi-infinite domain. Proc. Rom. Acad., Ser. A **16**, 490-498 (2015)

21. Doha, EH, Bhrawy, AH, Baleanu, D, Ezz-Eldien, SS, Hafez, RM: An efficient numerical scheme based on the shifted orthonormal Jacobi polynomials for solving fractional optimal control problems. Adv. Differ. Equ. **2015**, 15 (2015)

22. El Raheem, ZF, Salman, SM: On a discretization process of fractional-order logistic differential equation. J. Egypt. Math. Soc. **22**, 407-412 (2014)

23. El-Sayed, AMA, El-Raheem, ZF, Salman, SM: Discretization of forced Duffing system with fractional-order damping. Adv. Differ. Equ. **2014**, 66 (2014)

24. El-Sayed, AMA, Salman, SM: On a discretization process of fractional-order Riccati differential equation. J. Fract. Calc. Appl. **4**(2), 251-259 (2013)

25. Elaidy, SN: An Introduction to Difference Equations, 3rd edn. Undergraduate Texts in Mathematics. Springer, New York (2005)

26. Palis, J, Takens, F: Hyperbolicity and Sensitive Chaotic Dynamics and Homoclinic Bifurcation. Cambridge University Press, Cambridge (1993)

27. Puu, T: Attractors, Bifurcation and Chaos: Nonlinear Phenomena in Economics. Springer, Berlin (2000)

28. Udwadia, FE, von Bremen, H: A note on the computation of the largest *p*-Lyapunov characteristic exponents for nonlinear dynamical systems. J. Appl. Math. Comput. **114**, 205-214 (2000)

CHAPTER 11

Expansivity and Cone-fields in Metric Spaces

Łukasz Struski, Jacek Tabor

Faculty of Mathematics and Computer Science, Łojasiewicza 6, 30-348 Kraków, Poland

ABSTRACT

Due to the results of Lewowicz and Tolosa expansivity can be characterized with the aid of Lyapunov function. In this paper we study a similar problem for uniform expansivity and show that it can be described using generalized cone-fields on metric spaces. We say that a function f:X→X is uniformly expansive on a set Λ⊂X if there exist ε>0 and α∈(0,1) such that for any two orbits x:{−N,…,N}→Λ, v:{−N,…,N}→ of ffwe have

$$\sup_{-N \leq n \leq N} d(x_n, v_n) \leq \varepsilon \implies d(x_0, v_0) \leq \alpha \sup_{-N \leq n \leq N} d(x_n, v_n).$$

It occurs that a function is uniformly expansive if there exists a generalized cone-field on X such that ff is cone-hyperbolic.

KEYWORDS

Cone-field, Hyperbolicity, Expansive map, Lyapunov function

1. INTRODUCTION

In 1892 Lyapunov [9] introduced the idea of Lyapunov functions to study stability of equilibria of differential equations. The Lyapunov approach allows to assess the stability of equilibrium points of a system without solving the differential equations that describe the system. This theory is widely used in qualitative theory of dynamical systems.

In Lewowicz [7, 8]proposed to use Lyapunov functions of two variables to study structural stability and similar concepts, such as topological stability and persistence. The method has been applied in particular to study hyperbolic diffeomorphisms on manifolds. For the survey of the results, methods and possible generalizations see [12].

Let us quote one of the most interesting results from [12]. Let f:M→M be a homeomorphism of a compact manifold MM. For U:M×M→R we define

$$\Delta_f U(x, y) := U(f(x), f(y)) - U(x, y) \text{ for } x, y \in M.$$

We say that UU is a Lyapunov function for ff if it is continuous, vanishes on the diagonal, and ΔfU(x,y) is positive for (x,y) on a neighborhood of the diagonal, x≠y.

The following result characterizes expansive homeomorphisms in terms of Lyapunov functions.

Theorem

[12, Theorem 3.2]. Let ff be a homeomorphism of a compact manifold MM. The following conditions are equivalent:

a) ff is expansive;

b) there exists a Lyapunov function for ff.

The proof of this result for diffeomorphisms ff can be found in [7]; see Sect. 4 and Lemma 3.3 of that paper. Additional arguments required for the case of a homeomorphism are discussed in [6], Sect. 1]. See also [12], where Tolosa, basing on the results of Lewowicz, characterized the expansivity on metric spaces with the using Lyapunov functions.

In this paper we use a generalized notion of cone-fields on metric space to describe uniform expansivity. The notions of cone-fields and cone condition [4, 10] originally appeared in the late 60's in the works of Alekseev, Anosov, Moser and Sinai. Recently, Sheldon Newhouse [10] obtained new conditions for dominated and hyperbolic splittings on compact invariant sets with the use of cone-fields. It is also worth mentioning that the notion of cone-field can be very useful in the study of hyperbolicity [1, 3, 4, 10].

Let us briefly describe the contents of this paper. In Sect. 2 we discuss the notion of uniform expansivity. We show that if f is uniformly expansive then it is also expansive. In Sect. 3 we recall our generalization of cone-fields to metric spaces which we presented in paper [11] and show that the existence of hyperbolic cone fields guarantees uniform expansivity. In Sect. 4 we show how to construct functions cscs, cucu for a uniformly expansiveff such that ff is cone-hyperbolic with respect to the cone-field (cs,cu)(cs,cu). The main result of the section can be summarized as follows:

Main Result

[see Theorem 3]. Let XX be a metric space and let f:X⇀X be an L-bilipschitz map. Assume that Λ⊂X is an invariant set for f such that f is uniformly expansive on Λ.

Then there exists a cone-field on ΛΛ such that f is cone-hyperbolic on Λ.

2. UNIFORM EXPANSIVITY

First we define uniform expansivity of ff and show that this notion is stronger than the classical expansivity.

By a partial map from X to Y (written as f:X⇀Y) we denote a function which domain is subset of X [2, Chapter 2]. By dom(f) we denote the domain of a partial map f:X⇀Y, and by im(f)im(f) we denote its inverse image. For a given f:X⇀X we say that a sequence x:I→X defined on a subinterval[1] I of Z is an *orbit of* f if

$$x_n \in \mathrm{dom}(f) \text{ and } x_{n+1} = f(x_n) \text{ for } n \in I \text{ such that } n+1 \in I.$$

We recall the classical definition of expansivity. We say that f:X⇀X is *expansive* on Λ⊂X if there exists an ε>0 such that for any two orbits x:Z→Λ, v: if $\sup d(x_n, v_n) \leq \varepsilon$ then $x = v$.

Definition 1

Let N∈N, ε>0 and α∈(0,1) be given. We say that f:X⇀X is (N,ε,α) *-uniformly expansive on a set* Λ⊂X if for any two orbits x:{−N,…,N}→Λ, v:{−N,…,N}→X we have

$$d_{\sup}(x, v) \leq \varepsilon \implies d(x_0, v_0) \leq \alpha d_{\sup}(x, v),$$

where

$$d_{\sup}(x, v) := \sup_{-N \leq n \leq N} d(x_n, v_n).$$

This notion is more useful because it does not need an infinite trajectory.

Example 1

Consider a rotation of f:S1→S1 by an angle αα. Then f is an isometry, and therefore is not expansive, and consequently not (N,ε,α) uniformly expansive on Λ=S1.

Example 2

Let us consider the function $f : \mathbb{R}_+ \ni x \mapsto x + \sqrt{x} \in \mathbb{R}_+$. One can easily check that this function is expansive because its derivative at each point is strongly greater than 11. On the other hand, ff is not uniformly expansive because for sufficiently large xx the derivative of the function at xx can become as close to 11 as we want.

One can easily verify that uniform expansivity implies classical expansivity (this result can also be easily deduced from Theorem 1 below).

Observation 1

[11, Observation 4.1] *Let $N \in \mathbb{N}$, $\varepsilon > 0$, $\alpha \in (0, 1)$, $\Lambda \subset X$ and $f : X \rightharpoonup X$* be given. If f is(N,ε,α)-uniformly expansive on Λ, then it is also expansive on Λ.

Given L≥1 and f:X⇀Y we call f L-*bilipschitz* if

$$L^{-1}d(x, y) \leq d(f(x), f(y)) \leq Ld(x, y) \text{ for } x, y \in \text{dom}(f). \tag{2.1}$$

Note that if a function f is L-bilipschitz then it is injective.

For δ>0and a set A⊂X we define the δ *-neighbourhood of* A as

$$A_\delta := \bigcup_{x \in A} B(x, \delta).$$

Let an injective map f:X⇀X be given. We call A⊂dom(f) an *invariant set for* $f(x)$ and $f^{-1}(x) \in A$ for every $x \in A$.

Now we show how to change the metric so that the function ff which is (N,·,·)-uniformly expansive becomes (1,·,·)-uniformly expansive.

Theorem 1

Let f:X⇀X be an L-bilipschitz map for some L>1 and α∈(0,1). Let Λ⊂X and δ>0 be such that Λδ⊂dom(f)∩im(f) We assume that ΛΛ is an invariant set for f and that f is (N,δ,α) -uniformly expansive on Λ.

Then there exists a metric ρρ on $\Lambda_{\delta L^{-N+1}}$ such that

$$d(x, v) \leq \rho(x, v) \leq L^{N-1} d(x, v) \ \ for \ \ x, v \in \Lambda_{\delta L^{-N+1}}, \tag{2.2}$$

that $f\ is\ (1, \delta L^{-N+1}, \sqrt[N]{\alpha})$- uniformly expansive on $\Lambda_{\delta L^{-N+1}}$ *and* $\max\{\alpha^{-1/N}, L\}$- bilipschitz map with respect to the metric ρρ.

Proof

Let $\beta = \sqrt[N]{\alpha}$. We put

$$\rho(x, v) := \max_{k \in \{-N+1, \dots, N-1\}} \beta^{|k|} d(f^k(x), f^k(v)) \ \text{ for } x, v \in \Lambda_{\delta L^{-N+1}}.$$

Inequalities (2.2) follow from the definition and (2.1). Note that for $k \in \{-N+1, \ldots, N-1\}$ } we have

$$x, v \in \Lambda_{\delta L^{-N+1}} \Longrightarrow f^k(x), f^k(v) \in \Lambda_{\delta L^{-N+1+|k|}}.$$

This means that ρρ is well defined on $\Lambda_{\delta L^{-N+1}}$.

First we show that ff is max $\{\beta^{-1}, L\}$-bilipschitz map with respect to the metric ρ. Since ffis LL-bilipschitz in the metric dd, we know that $d(f^N(x), f^N(v)) \leq Ld(f^{N-1}(x), f^{N-1}(v))$ and finally we get

$$\begin{aligned}
\rho(f(x), f(v)) &= \max_{k \in \{-N+1, \ldots, N-1\}} \beta^{|k|} d(f^k(f(x)), f^k(f(v))) \\
&= \max\{\beta^{|-N+1|} d(f^{-N+2}(x), f^{-N+2}(v)), \ldots, \beta^{N-1} d(f^N(x), f^N(v))\} \\
&= \max\{\beta^{|-N+1|}\beta^{-1}\beta d(f^{-N+2}(x), f^{-N+2}(v)), \ldots, \beta^1\beta^{-1}\beta d(x, v), \\
&\quad \beta^0\beta\beta^{-1} d(f(x), f(v)), \ldots, \beta^{N-2}\beta\beta^{-1} d(f^{N-1}(x), f^{N-1}(v)), \\
&\quad \beta^{N-1} d(f^N(x), f^N(v))\} \\
&= \max\{\beta\beta^{|-N+2|} d(f^{-N+2}(x), f^{-N+2}(v)), \ldots, \beta\beta^0 d(x, v), \\
&\quad \beta^{-1}\beta^1 d(f(x), f(v)), \ldots, \beta^{-1}\beta^{N-1} d(f^{N-1}(x), f^{N-1}(v)), \\
&\quad \beta^{N-1} d(f^N(x), f^N(v))\} \\
&\leq \max\{\beta\beta^{|-N+2|} d(f^{-N+2}(x), f^{-N+2}(v)), \ldots, \beta\beta^0 d(x, v), \\
&\quad \beta^{-1}\beta^1 d(f(x), f(v)), \ldots, \beta^{-1}\beta^{N-1} d(f^{N-1}(x), f^{N-1}(v)), \\
&\quad \beta^{N-1} L d(f^{N-1}(x), f^{N-1}(v))\} \\
&\leq \max\{\beta, \beta^{-1}, L\} \cdot \rho(x, v) = \max\{\beta^{-1}, L\} \cdot \rho(x, v).
\end{aligned}$$

Similarly, as for the opposite inequality, we know that $L^{-1} d(f^{N-1}(x), f^{N-1}(v)) \leq$

$d(f^N(x), f^N(v))$ and $L^{-1} d(f^{-N+1}(x), f^{-N+1}(v)) \leq d(f^{-N+2}(x), f^{-N+2}(v))$. Hence

$$\begin{aligned}
\rho(f(x), f(v)) &= \max\{\beta\beta^{|-N+2|} d(f^{-N+2}(x), f^{-N+2}(v)), \ldots, \beta\beta^0 d(x, v), \\
&\quad \beta^{-1}\beta^1 d(f(x), f(v)), \ldots, \beta^{-1}\beta^{N-1} d(f^{N-1}(x), f^{N-1}(v)), \\
&\quad \beta^{N-1} d(f^N(x), f^N(v))\} \\
&\geq \max\{\beta\beta^{|-N+2|} d(f^{-N+2}(x), f^{-N+2}(v)), \ldots, \beta\beta^0 d(x, v), \\
&\quad \beta^{-1}\beta^1 d(f(x), f(v)), \ldots, \beta^{-1}\beta^{N-1} d(f^{N-1}(x), f^{N-1}(v)), \\
&\quad \beta^{N-1} L^{-1} d(f^{N-1}(x), f^{N-1}(v))\} \\
&\geq \min\{\beta, L^{-1}\} \cdot \rho(x, v).
\end{aligned}$$

Now we show that for for $x \in A$ and $v \in A_{\delta L^{-N+1}}$ such that

$$\max\left\{\rho(f^{-1}(x), f^{-1}(v)), \rho(x, v), \rho(f(x), f(v))\right\} \leq \delta L^{-N+1} \tag{2.3}$$

the following inequality holds:

$$\rho(x, v) \leq \beta \max(\rho(f(x), f(v)), \rho(f^{-1}(x), f^{-1}(v))).$$

We have to show that for k=–N+1,…,N–1

$$\beta^{|k|} d(f^k(x), f^k(v)) \leq \beta \max \times \left(\max_{k=-N+1,\ldots,N-1} \beta^{|k|} d(f^{k+1}(x), f^{k+1}(v)), \max_{k=-N+1,\ldots,N-1} \beta^{|k|} d(f^{k-1}(x), f^{k-1}(v))\right).$$

For k<0 or k>0 it is straightforward. Consider the case k=0. From (2.2) and (2.3) we get

$$\max\left\{d(f^{-1}(x), f^{-1}(v)), d(x, v), d(f(x), f(v))\right\} \leq \delta L^{-N+1},$$

which together with (2.1) implies that $d(f^k(x), f^k(v)) \leq \delta$ for $k = -N, \ldots, N$. By the uniform expansivity and the fact that β<1 we get

$$d(x, v) \leq \alpha \max_{|k| \leq N} d(f^k(x), f^k(v)) \leq \beta \max_{|k| \leq N} (\beta^{N-1} d(f^k(x), f^k(v)))$$
$$\leq \beta \max\left(\max_{|k| \leq N-1} \beta^{|k|} d(f^{k+1}(x), f^{k+1}(v)), \max_{|k| \leq N-1} \beta^{|k|} d(f^{k-1}(x), f^{k-1}(v))\right).$$

3. CONE-FIELDS AND CONE-HYPERBOLIC MAPS

In this section, for the convenience of the reader, we recall basic definitions concerning generalization of cone-fields to metric spaces (for more information and motivation see [5,11]).

Definition 2

[11, Definition 3.1] Let δ>0 and Λ⊂X be nonempty. We say that a pair of functions cs,cu:U→R+cs,cu: for some U⊂X×X forms a δ -*cone-field on* ΛΛ if

$$\{x\} \times B(x, \delta) \subset U \quad \text{for } x \in \Lambda.$$

We put $c(x, v) := \max\{c_s(x, v), c_u(x, v)\}$. If there exists K>0 such that:

$$\frac{1}{K} d(x, v) \leq c(x, v) \leq K d(x, v) \text{ for } (x, v) \in U$$

then we call it (K,δ) *-cone-field on* Λ or *uniform* δ *-cone-field on* Λ.

For each point x∈Λ we introduce *unstable* and *stable cones* by the formula

$$C_x^u(\delta) := \{v \in B(x, \delta) \; : \; c_s(x, v) \leq c_u(x, v)\},$$
$$C_x^s(\delta) := \{v \in B(x, \delta) \; : \; c_s(x, v) \geq c_u(x, v)\}.$$

We consider a partial map f:X⇀Y between metric spaces X and Y and Λ⊂dom(f) . Assume that X is equipped with a uniform δ-cone-field on Λ and Y is equipped with a uniform δδ-cone-field on a subset Z of Y such that f(Λ)⊂Z.

For every x∈dom(f) we put

$$B_f(x,\delta) := \{v \in B(x,\delta) \cap \mathrm{dom}(f) : f(v) \in B(f(x),\delta)\}.$$

Now we define ux(f;δ) and sx(f;δ) the expansion and the contraction rates of ff, respectively. These rates are a modification of the classical definition from [10], but we do not assume that the function ff is invertible (for more information see [11]).

Definition 3

[11, Definition 3.2] Let x∈dom(f) and δ>0 be given. We define

$$u_x(f;\delta) := \sup\{R \in [0,\infty] \mid c(f(x), f(v)) \geq Rc(x,v), v \in B_f(x,\delta); v \in C_x^u(\delta)\},$$
$$s_x(f;\delta) := \inf\{R \in [0,\infty] \mid c(f(x), f(v)) \leq Rc(x,v), v \in B_f(x,\delta); f(v) \in C_{f(x)}^s(\delta)\}.$$

Let $u_\Lambda(f;\delta) := \inf\limits_{x\in\Lambda}\{u_x(f;\delta)\}$ and $s_\Lambda(f;\delta) := \sup\limits_{x\in\Lambda}\{s_x(f;\delta)\}$.

Definition 4

We say that f is δ *-cone-hyperbolic* on Λ if

$$s_\Lambda(f;\delta) < 1 < u_\Lambda(f;\delta).$$

The next proposition is a simple analogue of [10, Lemma 1.1].

Proposition 1

[11, Proposition 3.1] Every δδ-cone-hyperbolic is δδ-cone-invariant, i.e. for x∈Λ and v∈Bf(x,δ) we have

$$v \in C_x^u(\delta) \implies f(v) \in C_{f(x)}^u(\delta),$$

and

$$f(v) \in C^s_{f(x)}(\delta) \implies v \in C^s_x(\delta).$$

Theorem 2

[11, Theorem 4.1] Suppose that for K>0 and δ>0 we are given a (K,δ)-cone-field on Λ⊂X . Let f:Λδ→X be δ -cone-hyperbolic on Λ and let λ>1 be chosen such that

$$s_\Lambda(f;\delta) \leq \lambda^{-1},\ u_\Lambda(f;\delta) \geq \lambda.$$

Then f is $(N, \delta, K^2/\lambda^N)$)-uniformly expansive on ΛΛ for every $N \in \mathbb{N},\ N > 2\log_\lambda K$.

Example 3

Let $f: T^2 \to T^2$ be defined by $f(x, y) = (2x + y, x + y)$, where $T^2 = \mathbb{R}^2/\mathbb{Z}^2$.

We know that ff is expansive (see [4, Sect. 1.8]). It is easy to show that

$$s_{T^2}(f;\delta) \leq \frac{3-\sqrt{5}}{2} < 1, \quad u_{T^2}(f;\delta) \geq \frac{3+\sqrt{5}}{2} > 1.$$

From Theorem 2 we conclude that ff is uniformly expansive on Λ=T2.

4. EXPANSIVITY AND CONE-FIELDS

In this section we show that uniform expansiveness of ff on an invariant set Λ lets us construct a cone-field on ΛΛ such that ff is cone-hyperbolic on Λ. In our reasoning we will need the notion of εε-quasiconvexity.

Definition 5

Let II be a subinterval of Z, and let ε≥ be fixed. We call a sequence α:I→Rεα: *-quasiconvex* if

$$\alpha_n \leq \max\{\alpha_{n-1}, \alpha_{n+1}\} - \varepsilon \text{ for } n \in I : n-1, n+1 \in I.$$

Now we show some properties of εε-quasiconvex sequences, which will be used later.

Observation 2

Let $\varepsilon \geq 0$ *and* $\alpha : I \to \mathbb{R}$ be an εε-quasiconvex sequence.

Then

i) if $m, m+2 \in I$ *and* $\alpha_{m+1} > \alpha_m - \varepsilon$ *then*

$$\alpha_{n+1} \geq \alpha_n + \varepsilon \textit{ for } n \geq m+1 \textit{ such that } n, n+1 \in I. \tag{4.1}$$

ii) $m-1, m+1 \in I$ *and* $\alpha_{m+1} < \alpha_m + \varepsilon$ *then*

$$\alpha_{n+1} \leq \alpha_n - \varepsilon \textit{ for } n < m \textit{ such that } n, n+1 \in I. \tag{4.2}$$

Proof

The above statements are similar so we show the first one. The proof proceeds on induction. Suppose that $m+2 \in I$ and $\alpha_{m+1} > \alpha_m - \varepsilon.$. Since α is ε-quasiconvex,

$$\alpha_{m+1} \leq \max\{\alpha_m, \alpha_{m+2}\} - \varepsilon = \max\{\alpha_m - \varepsilon, \alpha_{m+2} - \varepsilon\}.$$

But $\alpha_{m+1} > \alpha_m - \varepsilon,$ so we get

$$\alpha_{m+1} \leq \alpha_{m+2} - \varepsilon,$$

and hence

$$\alpha_{m+2} \geq \alpha_{m+1} + \varepsilon.$$

It implies that (4.1) is valid for n=m+1. Suppose now that (4.1) holds for some n≥m+1, i.e. that n;n+1∈In; and αn+1≥αn+ε. Assume additionally that n+2∈I. Then we get

$$\alpha_{n+1} \leq \alpha_{n+2} - \varepsilon,$$

thus

$$\alpha_{n+2} \geq \alpha_{n+1} + \varepsilon,$$

which completes the proof.

The following proposition will be a basic tool in the proof of our main result, Theorem 3.

Proposition 2

Let $\varepsilon > 0, L > 1, \beta \in (0, 1)$ *and let* (Y, ρ) be a metric space. *Let* $\Lambda \subset Y$ *be* be given and f:Y⇀Y be an L-bilipschitz map such that $\Lambda_\varepsilon \subset dom(f) \cap im(f)$. Assume that ΛΛ is an invariant set for ff and that f *is* $(1, \varepsilon, \beta)$ uniformly expansive on ΛΛ.

Then

$$c_s(x,v) := \inf\{\rho(f^k(x), f^k(v)) \mid k \in (-\infty, 0) \cap \mathbb{Z} : f^l(v) \in B(f^l(x), \varepsilon)$$
$$\text{for } l \in [k, 0] \cap \mathbb{Z}\},$$
$$c_u(x,v) := \inf\{\rho(f^k(x), f^k(v)) \mid k \in [0, \infty) \cap \mathbb{Z} : f^l(v) \in B(f^l(x), \varepsilon)$$
$$\text{for } l \in [0, k] \cap \mathbb{Z}\}, \tag{4.3}$$

define an (L,ε/L) cone-field on Λ. Moreover, f is cone-hyperbolic on Λ and

$$s_\Lambda(f; \varepsilon/L) \le \beta < \frac{1}{\beta} \le u_\Lambda(f; \varepsilon/L). \tag{4.4}$$

Proof

First we show that $c_s(x,v)$ and $c_u(x,v)$ defined above are $(L, \varepsilon/L)$ cone-field on Λ, i.e.

$$\frac{1}{L}\rho(x,v) \le c(x,v) \le L\rho(x,v) \text{ for } (x,v) \in \{(x,v) : x \in \Lambda, v \in B(x, \varepsilon/L)\},$$

where $c(x,v) := \max\{c_s(x,v), c_u(x,v)\}$.

Choose an arbitrary point $x \in \Lambda$ and $v \in B(x, \varepsilon/L)$.. We can assume that x≠v, because the case $x = v$ is trivial $(c_s(x,v) = c_u(x,v) = 0 = \rho(x,v))$.

Let II be the biggest subinterval of Z containing 0 such that

$$\sup\{\rho(f^n(x), f^n(v)) : n \in I\} \le \varepsilon. \tag{4.5}$$

Since ff is LL-bilipschitz, we know that $f^{-1}(v) \in B(f^{-1}(x), \varepsilon)$, and therefore $\{-1, 0\} \subset I$. This yields $c(x,v) < \infty$.

Now we define a sequence $\{a_n\}_{n \in I} \subset \mathbb{R}$ by the formula

$$a_n := \ln \rho(f^n(x), f^n(v)) \text{ for } n \in I. \tag{4.6}$$

Observe that anan is well-defined because $\rho(f^n(x), f^n(v)) > 0$ for all $n \in I$..

Let

$$I_- := \{n \in I : n < 0\} \text{ and } I_+ := \{n \in I : n \geq 0\}.$$

We have the following relations:

$$c_s(x, v) = \exp\left(\inf_{n \in I_-} \{a_n\}\right) \text{ and } c_u(x, v) = \exp\left(\inf_{n \in I_+} \{a_n\}\right).$$

where we use the convention $\exp(-\infty) = 0.$.

We show that the sequence $\{a_n\}$ is $\ln(1/\beta)$-quasiconvex. Let $n \in I$ be such that n−1, n+1∈I. By (4.5) we observe that

$$\max\{\rho(f^{n-1}(x), f^{n-1}(v)), \rho(f^n(x), f^n(v)), \rho(f^{n+1}(x), f^{n+1}(v))\} \leq \varepsilon.$$

Consequently, by (1,ε,β) uniform expansivity of f we get

$$\rho(f^n(x), f^n(v)) \leq \beta \max\{\rho(f^{n-1}(x), f^{n-1}(v)), \rho(f^{n+1}(x), f^{n+1}(v))\},$$

which implies that $a_n \leq \max\{a_{n-1}, a_{n+1}\} - \ln(1/\beta).$.

Now we consider two cases. If a−1≤a then by Observation 2 i) we get

$$a_{n+1} \geq a_n + \ln\frac{1}{\beta} \text{ for } n \geq 0, n \in I,$$

which yields

$$\inf_{n\in I_-}\{a_n\} \le a_{-1} \le a_0 = \inf_{n\in I_+}\{a_n\},$$

Hence

$$c_s(x,v) \le c_u(x,v) = c(x,v) = e^{a_0} = \rho(x,v).$$

On the other hand if a−1≥a0 then by Observation 2 ii) we get

$$a_{n+1} \le a_n - \ln\frac{1}{\beta} \text{ for } n < -1, n \in I.$$

Therefore

$$\inf_{n\in I_-}\{a_n\} = a_{-1} \ge a_0 \ge \inf_{n\in I_+}\{a_n\},$$

and consequently

$$c_u(x,v) \le c_s(x,v) = c(x,v) = e^{a_{-1}} = \rho(f^{-1}(x), f^{-1}(v)).$$

Since ff is L-bilipschitz, we get that cs,cucs,cu define an (L,ε/L) cone-field on Λ.

Now we check that ff is cone-hyperbolic on Λ. Let us take $x \in \Lambda$ and $v \in B_f(x, \varepsilon/L)$ such that $f(v) \in C^s_{f(x)}(\varepsilon/L)$. We define the sequence {an}n∈I as in (4.6).

We show that a0≥a1. Suppose that, on the contrary, a0<a1. By Observation 2 i) we get

$$a_{n+1} \ge a_n \text{ for } n \ge 1, n \in I.$$

Hence

$$\ln(c_u(f(x), f(v))) = \inf_{n \geq 1, n \in I} \{a_n\} = a_1 > a_0 \geq \inf_{n < 1, n \in I} \{a_n\} = \ln(c_s(f(x), f(v))),$$

which is a contradiction with $f(v) \in C^s_{f(x)}(\varepsilon/L)$. So we have $a_1 \leq a_0$. By the Observation 2 ii) we get

$$a_{n+1} \leq a_n - \ln(1/\beta) \text{ for } n < 0 \text{ such that } n, n+1 \in I.$$

In particular,

$$a_0 \leq a_{-1} - \ln(1/\beta). \tag{4.7}$$

Consequently,

$$c_u(f(x), f(v)) = \exp\left(\inf_{n \geq 1, n \in I} \{a_n\}\right) \leq \exp(a_1) \leq \exp(a_0)$$

$$= \exp\left(\inf_{n < 1, n \in I} \{a_n\}\right) = c_s(f(x), f(v)) = \underline{c(f(x), f(v))}$$

$$\overset{(4.7)}{\leq} \beta \exp(a_{-1}) = \beta \exp\left(\inf_{n \in I_-} \{a_n\}\right) \leq \underline{\beta c(x, v)}.$$

Therefore

$$s_A(f; \varepsilon/L) = \sup_{x \in A} \{s_x(f; \varepsilon/L)\} \leq \beta < 1.$$

Now we consider an $x \in A$ and $v \in B_f(x, \varepsilon/L)$ We show that $a_0 \geq a_{-1}$. Suppose the contrary, $a_0 \geq a_{-1}$. By Observation 2 ii) we get

$$a_{n+1} \geq a_n \text{ for } n < -1, n \in I.$$

Hence

$$\inf_{n\in I_-}\{a_n\} = a_{-1} > a_0 \geq \inf_{n\in I_+}\{a_n\},$$

which is contradiction with $v \in C_x^u(\varepsilon/L)$. So we have $a_0 \geq a_{-1}$. By the Observation 2 i) we get

$$a_{n+1} \geq a_n + \ln(1/\beta) \text{ for } n \geq 0 \text{ such that } n, n+1 \in I.$$

In particular,

$$a_1 \geq a_0 + \ln(1/\beta). \tag{4.8}$$

Finally

$$c_s(f(x), f(v)) = \exp\left(\inf_{n<1, n\in I}\{a_n\}\right) \leq \exp(a_0) \leq \exp(a_1)$$

$$= \exp\left(\inf_{n\geq 1, n\in I}\{a_n\}\right) = c_u(f(x), f(v)) = c(f(x), f(v)),$$

which yields

$$\exp(a_1) = \underline{c(f(x), f(v))} \overset{4.8}{\geq} \frac{1}{\beta}\exp(a_0) = \frac{1}{\beta}\exp\left(\inf_{n\in I_+}\{a_n\}\right) = \underline{\frac{1}{\beta}c(x, v)}.$$

This shows that

$$u_\Lambda(f; \varepsilon/L) = \inf_{x\in\Lambda}\{u_x(f; \varepsilon/L)\} \geq \frac{1}{\beta} > 1.$$

Therefore ff is cone-hyperbolic on Λ.

As a consequence of earlier results we obtain the following theorem.

Theorem 3

Let $\varepsilon > 0, L > 1, N \in \mathbb{N}, \alpha \in (0,1)$ be fixed. Let (X,d) be a metric space and Λ⊂X be given. Let f:X→X be an L-bilipschitz map such that $\Lambda_\varepsilon \subset dom(f) \cap im(f)$. Assume that Λ is an invariant set for f and that f is (N,ε,α)-uniformly expansive on Λ.

Then there exists an $(\max\{\alpha^{-1/N}L^{N-1}, L^N\}, \min\{\varepsilon L^{-2N+1}\sqrt[N]{\alpha}, \varepsilon L^{-2N}\})$, min cone-field on Λ such that f is cone-hyperbolic on ΛΛ and

$$s_\Lambda(f, \min\{\varepsilon L^{-2N+1}\sqrt[N]{\alpha}, \varepsilon L^{-2N}) \leq \sqrt[N]{\alpha} < \frac{1}{\sqrt[N]{\alpha}} \leq u_\Lambda(f, \min\{\varepsilon L^{-2N+1}\sqrt[N]{\alpha}, \varepsilon L^{-2N}).$$

Proof

We will apply Proposition 2. By applying Theorem 1 (for δ=ε) we obtain the metric ρ which is equivalent to dd on U={x:d(x,Λ)<εL−N+1} and such that

i) $d(x,v) \leq \rho(x,v) \leq L^{N-1}d(x,v)$ for $x, v \in U$,

ii) f is $(1, \varepsilon L^{-N+1}, \sqrt[N]{\alpha})$- uniformly expansive on UU with respect to the metric ρρ,

iii) f is $\max\{\alpha^{-1/N}, L\}$ bilipschitz map on U with respect to the metric ρ.

Let $\tilde{Y} = \{y : d(y,\Lambda) < L^{-N+1}\varepsilon\}$ and $\tilde{L} = \max\{\alpha^{-1/N}, L\}$. We use Proposition 2 $\tilde{\varepsilon} = \varepsilon L^{-N}, \tilde{L}, \tilde{\beta} = \sqrt[N]{\alpha}, \tilde{f} = f|_{\{x:d(x,\Lambda)<\varepsilon L^{-N}\}}$ and construct functions cs˜cu˜ which define an (L˜,δ˜) cone-field on UU such that f˜ is δ-cone-hyperbolic with respect to the metric ρρ, where δ˜=εL−N/L.

Now we need to "translate" the results from the metric ρρ to the original metric dd. For clarity of notation we use the subscript (.)d to denote objects with respect to the metric ddand (.)ρ(.)ρ to denote objects with respect to the metric ρρ.

By the definition of (L˜,δ˜) cone-field on UU and i) we get

$$\frac{1}{\widetilde{L}L^{N-1}}d(x,v) \le \frac{1}{\widetilde{L}}d(x,v) \le \frac{1}{\widetilde{L}}\rho(x,v) \le c(x,v) \le \widetilde{L}\rho(x,y) \\ \le \widetilde{L}L^{N-1}d(x,y) \text{ for } (x,v) \in \{x \in U, v \in B(x,\widetilde{\delta})_\rho\}.$$

From i) we have

$$B(x,\widetilde{\delta}/L^{N-1})_d \subset B(x,\widetilde{\delta})_\rho, \quad B_f(x,\widetilde{\delta}/L^{N-1})_d \subset B_f(x,\widetilde{\delta})_\rho,$$

and

$$C_x^u(\widetilde{\delta}/L^{N-1})_d \subset C_x^u(\widetilde{\delta})_\rho, \quad C_x^s(\widetilde{\delta}/L^{N-1})_d \subset C_x^s(\widetilde{\delta})_\rho.$$

Consequently, from Definition 3 for an arbitrary x∈U we get

$$u_x(f;\widetilde{\delta})_\rho \le u_x(f;\widetilde{\delta}/L^{N-1})_d, \quad s_x(f;\widetilde{\delta})_\rho \ge s_x(f;\widetilde{\delta}/L^{N-1})_d.$$

Hence

$$u_U(f;\widetilde{\delta})_\rho \le u_U(f;\widetilde{\delta}/L^{N-1})_d, \quad s_U(f;\widetilde{\delta})_\rho \ge s_U(f;\widetilde{\delta}/L^{N-1})_d.$$

From the above inequalities and (4.4)

$$s_U(f;\widetilde{\delta})_\rho \le \widetilde{\beta} < 1 < \frac{1}{\widetilde{\beta}} \le u_U(f;\widetilde{\delta})_\rho,$$

we obtain that f is $(\widetilde{\delta}/L^{N-1})$-cone-hyperbolic in metric dd and

$$s_U(f;\widetilde{\delta}/L^{N-1})_d \leq \widetilde{\beta} < 1 < \frac{1}{\widetilde{\beta}} \leq u_U(f;\widetilde{\delta}/L^{N-1})_d.$$

Finally we conclude that cs˜ and cu˜ are $(\max\{\alpha^{-1/N}L^{N-1}, L^N\}, \widetilde{\delta}/L^{N-1})$- - cone-field on Λ such that f is $(\widetilde{\delta}/L^{N-1})$- cone-hyperbolic on ΛΛ with respect to metric dd.\

FOOTNOTES

1 We say the II is a subinterval of Z if [k,l]∩Z⊂I for any k,l∈I.

REFERENCES

1. Capiński, M.J., Zgliczyński, P.: Cone conditions and covering relations for normally hyperbolic invariant manifolds. Discrete and Continuous Dynamical Systems A., 641–670, (2011)

2. Gallier, J.: Discrete Mathematics. Springer, New York (2011)

3. Hruska, S.: A numerical method for constructing the hyperbolic structure of comple Hénon mappings. Found. Comput. Math., 427–455, (2006)

4. Katok, A., Hasselblatt, B.: Introduction to the Modern Theory of Dynamical Systems. Cambridge University Press, Cambridge (1997)

5. Kułaga, T., Tabor, J.: Hyperbolic dynamics in graph-directed IFS. J. Differ. Equ. **251**, 3363–3380 (2011)

6. Lewowicz, J.: Expansive homeomorphisms of surfaces. Boletim da Soc. Bras. de Mat.-Bulletin/Braz. Math. Soc. **20**(1), 113–133 (1989)

7. Lewowicz, J.: Lyapunov functions and topological stability. J. Differ. Equ. **38**(2), 192–209 (1980)

8. Lewowicz, J.: Persistence of semi-trajectories. J. Dyn. Differ. Equ. **18**, 1095–1102 (2006)

9. Lyapunov, A.M.: Problème général de la stabilité du mouvement. Ann. Fac. Sci. Univ. Toulouse **9**, 203–475 (1907)

10. Newhouse, S.: Cone-fields, Domination, and Hyperbolicity, University in Modern Dynamical Systems and Applications. In: Brin, M., Hasselblatt, B., Pesin, Y. (eds.) pp. 419–433. Cambridge University Press, Cambridge (2004)

11. Struski, Ł., Tabor, J., Kułaga, T.: Cone-fields without constant orbit core dimension. Discret. Contin. Dyn. Syst.-Ser. A. **32**(10), 3651–3664 (2012)

12. Tolosa, J.: The method of Lyapunov functions of two variables. Contemp. Math. **440**, 243–271 (2007)

Index

G

H

I

K

L

M

N

O

P

R

S

T

V